McINTOSH

W9-BHT-594

LITTLE BOOK OF BIG IDEAS

Science

First published in the United States of America in 2007
by Chicago Review Press, Incorporated
814 North Franklin Street
Chicago, Illinois 60610

Conceived and produced by
Elwin Street Limited
144 Liverpool Road
London N1 1LA

Illustrations: Richard Burgess, Emma Farrarons
Designer: Thomas Keenes

ISBN-13: 978-1-55652-665-7
ISBN-10: 1-55652-665-2

Printed in China

LITTLE BOOK OF BIG IDEAS

Science

Dr. Peter Moore

CHICAGO
REVIEW
PRESS

Contents

Physics and Chemistry

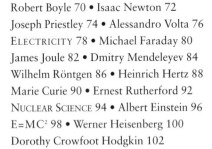

Mathematics

Introduction

As a way of thinking, science has produced remarkable results. Its method of attempting to make sense of the universe we live in by not only asking questions but going on to perform carefully organized data-generating experiments has let humans look inside minute atoms as well as probe the vastness of space.

It wasn't always this way. Ancient thinkers tried to work out how the world and the heavens worked by asking questions and then using processes of dialogue and argument to come to answers. Few thought that studying physical objects would provide anything useful and, until the 17th century, fewer still made any attempts at accurate measurements.

There are, however, always a few exceptions, and a few key individuals and groups living around the Mediterranean basin and the Middle East well before the Christian Era developed complex understandings of mathematics and used these to make some staggeringly detailed studies of the sun and stars.

Spanning two and a half thousand years of civilization, the *Little Book of Big Ideas: Science* gives you the stories behind the brilliance, brainstorming, setbacks, and formulas from cosmology and mathematics, biology and medicine, physics and chemistry. It lets you understand why these insights are of central importance in your life.

This book highlights fifty of the men and women who have made many of the most remarkable discoveries, and in each case focuses on one or two of the contributions they made. It sketches the person's background and shows how their initial ideas developed into tried-and-tested theories. From there you will see how these theories have long-lasting influence, affecting our lives and the technologies with which we live.

Through the book we discover how lines drawn in ancient sand influence the way that mathematicians today deal with trigonometry and the way 21st-century engineers design structures. We see how sequences of scientists and their discoveries led to ever fuller understanding of features such as the makeup of gases, how they eventually harnessed the power of electricity, and tackled the microscopic bugs that cause infection. In the realm of mathematics, you will see the wonder of curious numbers like pi, and may even develop a fresh appreciation of triangles.

The progress in science has been huge, the spin-offs immense. But this book shows that new ideas and insights are hard to predict, even though they may be just around the corner.

Dr. Peter Moore

Claudius Galen

We take it for granted that a doctor's knowledge of our bodies is more than skin deep, but this was not always the case. Claudius Galen was the person who established the importance of understanding how the body works.

Born: Ca. AD 129, Pergamum, Mysia (now Bergama, Turkey)
Education: Smyrna, Greece; Alexandria, Egypt
Major achievement: Pioneered human anatomy
Died: Ca. AD 200, place unknown.

As the Christian era dawned, the human body was seen predominantly as a vessel inhabited by a spirit. The vessel itself was of little interest — it was the spirit that demanded attention. Galen disagreed. He was fascinated by the physical structure of the body, and wanted to work out what each organ did.

Born to wealthy parents living in Pergamum, a thriving city on what is now the west coast of Turkey, Galen started studying medicine at the age of sixteen. At twenty he headed off to Smyrna in Greece to study anatomy with Pelops, one of the most respected physicians of the day. In his early thirties, Galen moved to Rome.

The Roman empire's love of blood sports, especially those involving human combat, gave Galen an extraordinary access to subjects. He spent much of his time working with soldiers injured in the Colosseum and he made a large number of key observations, many of which were related to blood.

> Employment is nature's physician, and is essential to human happiness.
>
> Claudius Galen

By being on hand to witness the effects of serious injury, he realized that two different types of blood came from deep cuts. One type was dark blue and ran sluggishly from cut vessels. These vessels had thin walls and by dissecting dead people, Galen found that he could trace these vessels back to the liver. As the liver was intricately connected to the gut, he concluded that food was broken down in the gut to form chyle, which was passed to the liver and there turned into blood. This then flowed out to the other organs of the body, where it was used up.

Galen also noted that one vessel went from the liver to the heart, and he was convinced that there were pores that allowed the blood to move between the chambers of the heart. His observations of animals and dead and dying soldiers revealed that the blood leaving the heart was bright red and seemed to be packed with life. It spurted from the vessel when it was cut and if allowed to flow, a person's life soon left his body. Galen's conclusion this time was that the heart packed the blood with vital spirit, the stuff of life, and that this pulsating fluid then distributed life throughout the body. In many ways these were good observations, but his conclusions were wrong and it took 1,700 years before William Harvey gained a better insight into blood and how it circulates around the body.

Galen's work fitted with his belief in a single god, which made it acceptable to the Christian and Islamic belief systems that dominated life in and around Europe, northern Africa, and the Middle East. As a consequence, his work was still revered until at least the 1600s.

Chyle: A milky fluid containing fat droplets which drains from the lacteals of the small intestine into the lymphatic system during digestion.

Anatomy

The study of human anatomy has a complex history, but over time anatomists have come to gain a significant insight into how the body works. This has radically affected the way doctors and surgeons try to cure disease or treat injury.

Whether you are comfortable with the idea of poking around inside dead bodies depends a lot on your view of the nature of life. Throughout history, some societies have seen it as a horrific intrusion into a person's being, while others have used it as the ultimate punishment for convicted murderers. Still others have looked inside bodies either as a means of fortune-telling, or of studying how the human body works.

The earliest records of anatomical studies are found on fragments of clay tablets made around 4000 BC in Nineveh, an ancient city in what is now northern Iraq. It appears that the temple priests made clay models of organs, such as the liver and lungs of sheep, and used observations of these organs for some form of fortune-telling.

A more developed understanding of anatomy can be seen in papyrus fragments written in about 1000 BC and found in Egypt. These show a relatively complex understanding of features such as our eyes, the digestive system, and bones. Much of this information probably came from the practice of mummifying bodies before burial.

According to Tertullian, a historian who lived in the ancient city of Carthage, in what is now Tunisia, sometime around AD 200, the study of human anatomy started to take off in Alexandria, Egypt. He says that Herophilus (ca. 335 BC–ca. 280 BC) and Erasistratus (born ca. 250 BC) vivisected up to 600 criminals. The Roman scholar Celsus (ca. 30 BC–AD 45) says that Herophilus obtained these criminals "for dissection alive, and contemplated, even while they breathed, those parts which nature had before concealed." All this work was lost, presumably in the great fire that destroyed the library of Alexandria in the early years of the Christian era.

A major step forward occurred when the Greek physician Claudius Galen (ca. AD 129–ca. AD 200) started studying anatomy, getting much of his insight while working with wounded gladiators in Rome. However, between Galen's observations and the late fifteenth century, very little new anatomical work was done. People felt that dissection was unethical and offensive on religious grounds, and Galen's work was simply taken as true and translated into many different languages. Then in 1490 the world-leading medical school at Padua in Italy opened a new anatomical theater. This stimulated people such as Leonardo da Vinci (1452–1519) and Andreas Vesalius (1514–1564) to start dissecting bodies again in an attempt to work out exactly what went on inside.

The seventeenth and eighteenth centuries saw the dawn of the Enlightenment in Europe, a golden era for science in which philosophers and scientists started to question previously held beliefs that had seen everything associated with life in terms of purely spiritual forces. Instead they looked for mechanical, verifiable explanations. Studying anatomy then became part of the process of seeing how the human machine functioned, a process that has continued ever since.

William Harvey

Carrying out experiments and assessing the data obtained are two key features of what we now think of as science, but for William Harvey they were revolutionary concepts. Through his experiments and the ensuing data, Harvey concluded that blood circulated around the body, an idea that led to a radical rethink of the nature of life.

Born: 1578, Folkestone, England
Education: University of Padua
Major achievement: Discovered that blood circulates around the body
Died: 1657, London, England

Seventeenth-century England was a place of turmoil and revolution. The country was divided by a civil war and savaged by the plague, while London was decimated by the Great Fire. At the same time, intellectuals were engaged in a revolution that was to have just as great an impact.

For centuries philosophers had ruled the world of thought, with the idea that you could deduce how something worked by thinking about it and judging your conclusions by seeing how they fit with the foundational work of great ancients such as Galen and Aristotle. But Harvey found that performing experiments could turn established ideas on their heads.

In one key set of experiments, Harvey concluded that blood circulated around the body. While Harvey had a great respect for the ancients, his sense of curiosity drove him to investigate for himself. Having been trained in Padua, Italy, at the world's foremost medical school, he returned to England. By nurturing an association with King Charles I, Harvey got permission to use the King's deer and other animals, which gave him the rare ability to perform experiments on a wide range of large, live animals.

His starting point for experiments on the heart was the ancient notion that systems in nature often operate in cycles. Seasons, day and night, the movement of the moon: all followed cyclical patterns. If that were so in the cosmos, it should also be so in the microcosm — the body. When he thought about blood he wondered if it too moved in a cycle.

So far he was operating as a classic philosopher, but then he did some practical experiments. By counting the number of times the heart beats, and making crude estimates of how much blood the heart could contain, he realized that it was far too much to be on a one-way journey to the rest of the body. By occluding the veins and arteries supplying and leaving the hearts of animals as diverse as deer and snakes, he saw that blood flowed into the heart from the veins and left it in the arteries.

In a leap of faith and imagination he then proposed that the blood passed from the arteries back into the veins within the organs. It took another half century before Italian Marcello Malpighi (1628–1694) saw capillaries through a primitive microscope and confirmed that Harvey's ideas were correct.

> [The heart] is the household divinity which, discharging its function, nourishes, cherishes, quickens the whole body, and is indeed the foundation of life, the source of all action.
>
> William Harvey

Harvey's work acts as a bridge between ancient philosophy and the mode of thinking that became known as science. He realized that to further medicine, you needed to make theories and then test them with experiments and measurements.

John Hunter

As a school dropout who hated books and studying and seemed only interested in collecting birds' nests, insects, and animals, there seemed little hope that John Hunter would make a name for himself, particularly in medicine. Born in Scotland, his professional break came when he moved to London at age twenty and worked as an assistant in his brother's anatomy school, a school that was to become increasingly famous for excellence.

Born: 1728, Long Calderwood, Scotland
Education: London
Major achievement: Pioneered surgery
Died: 1793, London, England

His first task was to dissect an arm to reveal its muscles, blood vessels, and nerves. Apparently the result was brilliant. Hunter became enthralled by the way that different animals had similar sets of organs. He studied anything he could lay his hands on, which in 1759 included a huge member of the dolphin family that had been caught in the mouth of the River Thames. It led him to some interesting conclusions, including one he noted in a textbook, where he wrote:

> The stomach is the distinguishing part between an animal and a vegetable, for we do not know any vegetable that has a stomach nor any animal without one.

In March 1761, at the height of the Seven Years War, Hunter traveled as an army surgeon to Belle-Île-en-Mer, a small island off the west coast of France. Here he studied anatomy as he tended soldiers who had been torn open in conflict. One aspect that fascinated him was the way that tissue became inflamed when it was damaged. He carefully recorded the way that blood vessels expand near to a wound, and believed that the stimulus driving

Above: An aneurysm is a localized weakness or injury to the wall of a blood vessel causing dilatation or ballooning and, in severe cases, causing hemorrhage or stroke. Can be caused by genetic predisposition or disease. From the Greek, *angeion* (blood vessel) and *eurys* (wide).

this was the same as the one that caused skin to change color when people blushed.

Hunter's passion for understanding how the body worked, coupled with a need to earn money, later caused him to become a surgeon. One operation that he performed drew directly from his knowledge of the way blood vessels worked. In December 1785 he operated on a forty-five-year-old coachman who for three years had had a painful swelling at the top of his leg, and could hardly walk. The swelling was caused by an aneurysm — a weakened artery that had blown up like a balloon and was not letting blood pass into the leg properly. Rather than amputate the leg, Hunter made an incision in the groin, over the site of the swelling. Knowing the anatomical arrangement of the muscles allowed him to part them carefully and reveal the damaged artery. He passed four lengths of tape under the swelling and tightened them so that the artery was restricted to its normal size. Within six weeks the coachman walked out of hospital.

Hunter combined his skills and curiosity and is now considered to have been not only a pioneer, but one of the greatest surgeons of all time.

Felix Hoffmann

Pain that stabs at the mind and screams for attention is the stuff of nightmares, and relief from pain has been a major focus of medicine throughout time. The modern solution is often to reach for a tablet of aspirin, but before 1897, when Felix Hoffmann synthesized the drug for the first time, the only options were ineffective analgesics or those with side effects that overshadowed their use.

Born: 1868, Ludwigsburg, Württemberg (modern-day Germany)
Education: University of Munich
Major achievement: Created aspirin
Died: 1946, Switzerland

Since antiquity, doctors such as Hippocrates had soaked the bark of willow trees and other plant materials to generate a "tea" that could relieve pain. Then in the early nineteenth century chemists isolated salicylic acid from these potions and discovered that it was the active ingredient. In 1859 German chemist Hermann Kolbe worked out the chemical structure of this acid and found a way of reconstructing it artificially in the laboratory, and in 1874 the Heyden Company, based just outside Dresden, began manufacturing salicylic acid as a painkiller.

The drug worked, but had unpleasant side effects; it irritated the stomach, and many people were simply unable to take it.

German research chemist Felix Hoffmann had a particular interest in pain relief because his father was afflicted by severe arthritis. As an employee of Bayer, a German chemical and pharmaceutical company, his job was to alter known compounds with the hope that these mutations could generate new drugs that had either greater potency, or fewer unwanted side effects. One of his favorite tricks was to add an acetyl group (CH_3CO) to existing molecules, and in August 1897 he added it to salicylic acid and generated acetylsalicylic acid. The head of Bayer's pharmaceutical laboratory, Heinrich Dreser, tested the substance on himself, set up a series of animal experiments, and then a batch of clinical trials. The results were clear: this compound was a hugely effective painkiller, with vastly reduced side effects.

Hoffman's discovery was developed, but Bayer's European patent application under the name "aspirin" — "a" from acetyl, and "spirin" from *Spiraea*, the Latin name for the genus of plants that naturally produce salicylic acid — was refused. A French chemist named Charles Gerhardt had already made an impure version of the compound without realizing its potential. They were, however, able to register an American patent, and started advertising worldwide. The result is that aspirin has become a household name, with around eighty billion tablets being taken each year.

It is intriguing to note that it took decades before scientists worked out *how* aspirin produced its effect. In 1982 John Vane won a Nobel Prize for his part in showing that aspirin blocks the action of COX-2, an enzyme that helps create prostaglandin, a hormone that triggers inflammation at the site of injury. As swelling often causes pain, removing inflammation reduces pain. We now realize that blocking this enzyme can help people with various forms of cancer, and can also reduce some people's risk of having a heart attack.

Hoffmann's desire to help his father has therefore led to huge benefits for countless millions of people.

Cancer: General term for diseases caused by abnormal and excessive growth and division of cells in the body.

Enzyme: A protein or protein-based molecule that speeds up chemical reactions in living things.

Hormone: A chemical messenger within the body that is secreted by one type of cell and acts on another type of cell.

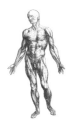

Karl Landsteiner

Taking blood from one person and infusing it into a patient seems to be a relatively straightforward process, but when it was first tried recipients often had huge reactions against the donated blood. When Karl Landsteiner's scientific work unraveled the mechanisms that enable safe blood transfusion he also opened the door for the development of successful organ transplants.

Born: 1868, Vienna, Austria
Education: University of Vienna
Major achievement: Discovered blood groups
Died: 1943, New York

Blood is composed of many different types of cell, each of which performs different duties. Red blood cells carry oxygen from lungs to tissues and enable blood to transport carbon dioxide from tissue to lungs. White cells fight infections, and platelets help blood to clot. These cells float in a liquid called plasma. If blood is allowed to clot, the plasma becomes serum — a liquid that has slightly different properties from plasma because the clotting factors have been used up.

In 1875, the German physiologist Leonard Landois took red cells from lambs' blood and mixed them with serum from a dog. He kept the mixture at body temperature and used a microscope to follow the action. By using serum he avoided any complication from clotting, but within two minutes the red blood cells burst. Landois postulated that this would also happen if you placed animals' blood in humans, and that this could lead to death. This was an important hypothesis. At that time some doctors were indeed infusing animals' blood into humans as an experimental therapy for various diseases. Those patients who did not die had severe adverse reactions, but doctors often took such fevers as signs that the patient was purging the disease from the body. Landois showed that it was the process itself that was damaging the blood. But what process made the two types of blood unable to mix? Twenty-five years later, Austrian scientist Karl Landsteiner started to answer the question.

He observed the way that blood from the twenty-two people working in his laboratory responded when it was mixed. Mixing serum from one person with whole blood taken from another sometimes caused the red cells to clump together. In other combinations this did not occur. Landsteiner concluded that people came from one of three groups, which he called A, B, and C. A year later, one of Landsteiner's pupils discovered a fourth group — a set of people whose serum didn't clump the red blood cells of either A- or B-type individuals. He called this group O.

> **Serum:** The clear, thin, and sticky fluid portion of the blood that remains after blood clots. Serum contains no blood cells, platelets, or fibrinogen.

This clumping process occurs because blood plasma contains a series of large molecules called antibodies. These search out proteins that come from outside the person's body — usually as a defense mechanism against bacteria and viruses. However, these antibodies will also attack injected foreign blood cells — if they can spot that they come from outside the body. Blood groups exist because blood cells are covered with tiny proteins, and while people from group A all have the same version of that protein, those in group B have a different version. If group B blood is injected into a group-A person, the recipient's antibodies lock on to these newly arrived proteins and trigger a response that destroys the cell. People in group O produce blood that has none of these marker proteins; such people are therefore "universal donors" — that is, their blood can be donated to any person. This amazing discovery has allowed us to go far beyond blood transfusion to complete organ transplant.

The Body

Carl Djerassi

Which came first, the oral contraceptive or the sexual revolution? Answering this question is difficult because the two became intrinsically intertwined. One thing is certain though: it was Carl Djerassi who on October 15, 1951, first synthesized the active ingredient of millions of tablets taken by women to prevent pregnancy.

Born: 1923, Vienna, Austria
Education: University of Wisconsin
Major achievement: Created the chemical behind the oral contraceptive
Died:

One of the ways of performing scientific experiments on animals is to make changes and see if they stop some natural process from occurring. Such changes often entail injecting a chemical, or making some physical alteration. If changes occur, you get an insight into the way the process is organized and controlled. With this in mind, Austrian professor of physiology Ludwig Haberlandt (1885–1932) took ovaries from a pregnant rabbit and implanted them into a non-pregnant rabbit. His aim was to see whether the ovaries released chemicals that controlled aspects of the rabbit's reproductive system. The results were clear: having an ovary from a pregnant rabbit prevented the recipient from becoming pregnant. The conclusion was equally clear: a chemical released by ovaries could prevent pregnancy.

Immediately Haberlandt spotted the implications. On January 20, 1927, a newspaper headline above a story reporting his work read, "My aim: fewer but fully desired children!" It took a few years, but scientists around the world succeeded in identifying the active agent in this process as progesterone, a hormone produced in large quantities by the ovaries of pregnant females. The problem now was that progesterone was difficult to purify from natural sources, hard to make artificially, and even then, it was poorly absorbed into the body if taken as a tablet.

The solution came from an inedible, wild species of South American yam. A researcher at Pennsylvania State College discovered that it produced vast quantities of diosgenin, a chemical that could be easily converted into progesterone. At this point Carl Djerassi stepped in and started synthesizing a range of different hormones from diosgenin. In 1951, working with a young Mexican chemist named Luis Miramontes, Djerassi synthesized 19-nor-17alpha-ethynyl-testosterone. Not the easiest name to swallow, so they called it norethindrone for short — but if taken as a tablet this steroid hormone was absorbed well and had all the power of progesterone.

Djerassi and his colleagues were excited about the product, but didn't see it as a world-altering compound. They were sure it would be useful in helping women with menstrual disorders, but its real impact became clear only after a group of medical researchers in Puerto Rico and Haiti ran a clinical trial in 1956 involving 6,000 women. When this showed that norethindrone could control the pregnancy rate, the world woke up to the realization that a drug had been found that could empower women, giving them more control over their reproductive destiny.

Soon "the Pill" was born. Though it might be unrealistic to say that norethindrone led the sexual revolution, it would be equally unrealistic to say that the sexual revolution would have occurred at the same rate without it. The Pill has radically changed the economic and social status of women and has improved women's health, by reducing unwanted pregnancies and miscarriages and also the incidence of ovarian and endometrial cancers, pelvic inflammatory disease, ovarian cysts, benign breast disease, iron deficiency anemia, and dysmenorrhea. Today, as many as 100 million women worldwide use the Pill.

Carl Linnaeus

By carefully studying plants and animals, Carl Linnaeus realized that you could place them in groups. However, his contribution to taxonomy goes far beyond contributing so-called scientific names to many of the world's plants and animals. Linnaeus developed a system of scientific classification whose grouping schemes now underlie many aspects of biology.

Born: 1707, Råshult, Sweden
Education: Universities of Lund and Uppsala in Sweden; University of Harderwijk, in the Netherlands
Major achievement: Developed classification for grouping animals
Died: 1778, Uppsala, Sweden

In 1731 and 1734, even before having completed his medical degree, Carl Linnaeus, or Carl von Linné as he was later called, mounted his first expeditions to collect and study plants, first in Lapland and then in central Sweden. By 1741 he held a professorial position at Sweden's prestigious University of Uppsala, where he specialized in treating people with syphilis. Although now tied down with teaching and medical duties, Linnaeus managed to send nineteen of his students on similar expeditions around the world.

As one of a group of eighteenth-century "natural theologians," Linnaeus had an underlying belief that since God had created the world, everything in it should be ordered and capable of being understood. The mechanism of sexual reproduction in plants had only just been discovered, and he studied the shapes and functions of reproductive organs from all the specimens that he and his students collected. This enabled Linnaeus to create a way of classifying each plant based on the number and shape of each one's reproductive organs — the male stamens, which produce pollen, and the female pistils, which collect the pollen and produce seeds.

As early as 1729 he wrote:

The flowers' leaves … serve as bridal beds which the Creator has so gloriously arranged, adorned with such noble bed curtains, and perfumed with so many soft scents …

Linnaeus' concentration on the sexual organs of plants produced some strange results, as for example it grouped fungi, mosses, lichens, algae, and ferns together, because they had no obvious sexual organs. Consequently it was criticized and later supplanted by systems that took into account many other physical features of a plant specimen.

One feature of Linnaeus's system that has survived is his binomial method that uses a two-part name for each species, e.g. *Homo sapiens*. The first part refers to the larger group, or "genus," of organisms, with the second part specifying an individual species within that group.

> A practical botanist will distinguish at the first glance the plant of the different quarters of the globe and yet will be at a loss to tell by what marks he detects them.
>
> Carl Linnaeus

This was similar to work first pioneered by Aristotle (384–322 BC), but Linnaeus took the idea further and grouped the genera into "orders," orders into "classes," and these into "kingdoms."

While this concept has been retained in biological study, new genetic findings at the end of the twentieth century have led biologists to switch some animals and plants between groups. Although the tables of names have changed, it nonetheless confirms that Linnaeus's overall concept was strong enough to be borne out over time.

Charles Darwin

When Charles Darwin concluded that new species had evolved from older ones by a process of natural selection, he created the theory of evolution and turned the world of biology, and in some ways theology, on its head.

Born: 1809, Shrewsbury, England
Education: University of Cambridge
Major achievement: Theory of evolution
Died: 1882, Downe, England

While studying at the University of Cambridge, Charles Darwin went on various field trips to examine geological formations and their fossils. He started to question whether, for example, birds had wings because a creator God wanted them to fly, or whether birds could fly *because* they had wings.

Darwin's big break came in 1831 when he set sail on HMS *Beagle* as the companion of the ship's twenty-six-year-old captain Robert FitzRoy (1805–1865). On the voyage they visited the Cape Verde Islands, the South American coast, the Strait of Magellan, the Galapagos Islands, Tahiti, New Zealand, Australia, the Maldives, and Mauritius, before returning to England in 1833.

While traveling, Darwin read Charles Lyell's (1797–1875) *Principles of Geology*, which argued that the world was continually being shaped and reshaped by ongoing geological forces. This ran against the accepted view that the world had been created a long time ago, and had only changed occasionally as a result of major natural events such as catastrophic floods.

Arriving in South America, Darwin found fossil evidence that seemed to indicate some form of progression from simple to complex life forms. In addition, on February 20, 1835, Darwin experienced an earthquake on the southwest coast of South America, which lifted the land by one to three meters. Later he found fossilized seashells high up in the mountains and wondered

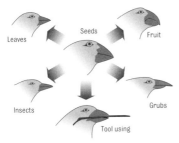

Left: On his groundbreaking trip to the Galapagos Islands, Darwin observed how the beaks of different species of finches were highly specialized depending on the food available on each island — a result, he argued, of natural selection.

Leaves

Seeds

Fruit

Insects

Grubs

Tool using

whether numerous previous quakes had driven them there. If so, this would support Lyell's theory that the earth was changing constantly.

From September to October 1835 Darwin visited the Galapagos Islands in the Pacific Ocean, which as far as he could see had been recently created by volcanic action. He was therefore surprized by the diversity of life that he found on the islands. On James Island he found seventy-one species of plants — of which thirty were unique to the island — and a further eight that were found only on other islands in this group. He was especially intrigued by the variety of species of finches, each with differently shaped beaks that seemed well suited to the particular food available on each island.

It turns out that Darwin wasn't alone in concluding that different species developed through a gradual process of evolution. Another explorer, Alfred Wallace (1823–1913) was studying hard in the Far East and had amassed over 125,000 specimens. Wallace sent a couple of letters to England outlining his ideas and, spurred on by this, Darwin quickly presented his theory to the Linnean Society of London on July 1, 1858. Just over a year later he published *On the Origin of Species*, one of the most influential books in the history of science.

Evolution

Charles Darwin's celebrated publication of *On the Origin of Species* forms just one step in a chain of events marking humanity's struggle to make sense of the biological world.

In ancient Greece, Aristotle (384–322 BC) and his student Theophrastus (later known as the father of botany) realized that they could place plants and animals into groups by comparing their physical features. This accompanied a realization that, on the whole, plants and animals could only produce offspring if they bred with other highly similar individuals — in other words, other members of the same species.

For over two thousand years the reason for this seemed obvious. God, or some creative influence, had generated all the species found on Earth and had made biological boundaries that ensured that species didn't get mixed up.

By the time that Charles Darwin (1809–1882) stepped aboard HMS *Beagle* and headed west to explore South America and the Galapagos Islands in the Pacific Ocean, people were already beginning to question this static idea of biological existence. Charles Lyell (1779–1875) had begun to show that looking at the structure of the Earth indicated that it may have existed for millions of years. This alone sparked questions of whether anything had changed significantly over that time.

On November 24, 1859, when Darwin eventually published his book, science took a massive step into the unknown — but a step that now underlies much of current biological thinking. The full title of his book reads *On the Origin of Species by Means of Natural Selection, or the Preservation of Favored Races in the Struggle for Life*. In it he shows how studying a wide range of plants and animals, as well as the environments they were found in, led him to conclude that animals which are best fitted to their environment are most likely to breed and pass on their characteristics to the next generation. Over enough time, he suggested, this process of gradual persistent change could generate individuals that were so different from their ancestors that they could no longer breed with non-changed members of the species. In other words, they developed to such an extent that they constituted a new species.

At the time Darwin was working, no one knew about DNA or genes and so the mechanism that allowed this passage of information from one generation to the next was a mystery. Austrian monk Gregor Mendel, however, was in the process of working out that physical characteristics were passed on in discrete units of information — what would become known as "genes" — and was making distinct progress in determining some of the mathematical and statistical rules that describe the way this happens.

> We can allow satellites, planets, suns, universe, nay whole systems of universes, to be governed by laws, but the smallest insect, we wish to be created at once by special act.
>
> Charles Darwin

The arrival of genetic technology now reveals more details of the underlying mechanisms. The discovery that most living organisms share similar basic housekeeping genes points to a common ancestry, and analysis of the differences within these genes gives an indication of what species have evolved from each other. Darwin would be fascinated.

Gregor Mendel

Although their lifespans overlapped, Gregor Mendel and Charles Darwin never met. But while Darwin was developing a theory of evolution, Mendel found statistical proof that plants and animals pass physical characteristics from one generation to the next.

Born: 1822, Heinzendorf, Austria
Education: University of Vienna
Major achievement: Realized that each physical characteristic was inherited separately
Died: 1884, Brünn, Austrian Empire (modern-day Brno, Czech Republic)

From the simple observation of livestock and crops, people knew that the offspring of plants or animals showed mixtures of their parents' physical characteristics. In his first paper, *Experiments in Plant Hybridization*, published in 1866, Mendel wrote that he wanted to discover a "generally applicable law of the formation and development of hybrids."

He described how he had carefully bred a type of pea, *Pisum sativum*. The plant was cheap and came in many strains that when bred in isolation produced identical offspring, but gave more complex results when mixed to produce hybrids.

Crossing a plant that produced smooth peas with another that produced wrinkled peas gave offspring that all had smooth peas. When these second-generation offspring were interbred, some of the new plants produced smooth, and some wrinkled peas. Intriguingly the ratio was almost perfectly three plants with smooth peas for every one with wrinkled peas.

He then looked at the way different colors were passed from one generation to the next and found that the same rule applied, and noted that inheritance of the color of both the plant's flowers and its peas occurred independently of pea shape.

Mendel studied with incredible diligence, cultivating some 28,000 pea plants to generate data for just two scientific papers. But it took genius to make sense of the results. Mendel realized

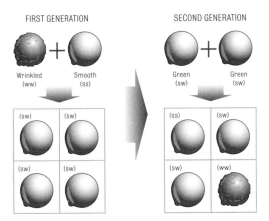

Wrinkled (ww) Smooth (ss)

Green (sw) Green (sw)

(sw) (sw)

(sw) (sw)

(ss) (sw)

(sw) (ww)

Above: As with eye color in humans, the texture of a pea plant's seeds is determined by a pair of genes. A plant will produce wrinkled peas only when *both* genes are of the recessive "wrinkled" variety (ww). Any other combination involving the dominant "smooth" gene (s) will result in a plant with smooth peas.

that such phenomena could only occur if three things were happening in the process of reproduction: first, characteristics must be carried from one generation to another by means of some physical "element" — we now call this element a "gene." Secondly, each characteristic must be recorded in the cells on a pair of these elements. Thirdly, Mendel concluded that some of the characteristics were dominant over others.

When Mendel died he was certain that he had made an important discovery and disappointed that no one seemed to care. Recognition only came in 1900 when three other scientists, Dutchman Hugo de Vries (1848–1935), German Carl Correns (1864–1933), and Austrian Erik Tschermak (1871–1962), read Mendel's work. They had each unknowingly repeated his work and discovered similar rules — but Mendel had got there first.

Biological Systems

Barbara McClintock

Before scientists knew the structure of DNA, and while they still had no idea how it carried information, Barbara McClintock's theory that genes could break free and move around within the chromosomes set genetics on a new path of discovery.

Born: 1902, Hartford
Education: Cornell
University, Botany
Major achievement:
Discovered that some genes
are able to move around the
chromosomes
Died: 1992, Huntington, New
York

As Barbara McClintock started her research, scientists thought that the dumbbell-shaped features inside each cell's nucleus probably carried biological information, and that these "chromosomes" somehow managed to pass the information to new individual cells during the process of cell division. Scientists also believed that traits such as the color of a flower and the shape of a pea could be inherited separately, because during the cell division that generates pollen and seeds (or sperm and eggs in animals), parts of the various chromosomes physically broke off and swapped with each other. It was a suggestion, but there was no evidence to support it.

Shortly after receiving her doctorate in 1927, McClintock began work with Harriet Creighton (1909–2004) at Cornell University. McClintock started by working out ways of identifying the ten different chromosomes present in maize cells. Then, working with a particular strain of maize, *Zea Mays*, that had a mutated chromosome number nine, the two researchers showed that parts of the chromosome did indeed swap over during the process that generates reproductive cells.

Next, working with Lewis Stadler (1869–1954) at the University of Missouri, McClintock started studying maize that had been exposed to X-rays. Peering into their cells with her microscope, she identified ring chromosomes. These she correctly suggested were parts of chromosomes that had been broken by radiation and subsequently fused to form a ring. Chromosomes occur in cells in pairs, and McClintock saw that chromosomes could break and fuse with the other member of the pair, before being ripped apart when the cell divided — a phenomenon that became known as the breakage-fusion-bridge cycle.

Moving to Cold Spring Harbor, McClintock discovered a bizarre genetic behavior in some of her breakage-fusion-bridge strains. Certain genes appeared to move from cell to cell during development of the corn kernel. When she presented this data at a meeting in 1951 she expected recognition and acceptance, but instead was greeted with silence and some derision. As she said in her Nobel prize acceptance speech, "Because I became actively involved in the subject of genetics only twenty-one years after the rediscovery, in 1900, of Mendel's principles of heredity ... acceptance of these principles was not general among biologists."

The problem was that many of the powerful people in science backed a theory of genetics that suggested that the function of a gene depended on where it was in a chromosome. If that were the case, then genes just couldn't jump around; or if they did, they would cease to function. McClintock's data only made sense if genes were distinct units that could move around on their own but still function. In time McClintock was proved right, and her theory of the nature of genes underpins current genetics.

Gene: The segment of DNA on a chromosome that contains the information necessary to make a particular protein.

Chromosome: A threadlike package of genes in the nucleus of a cell, made of DNA wrapped around supporting proteins.

Crick and Watson

The names Crick and Watson are so firmly linked together that it almost seems they are one person. Their joint fame rests on their momentous announcement, in 1953, of their discovery of the structure of DNA.

Francis Crick
Born: 1916, Northampton, England
Education: University College London
Major achievement: Discovered the structure of DNA
Died: 2004, San Diego

James Watson
Born: 1928, Chicago
Education: University of Chicago
Major achievement: Discovered the structure of DNA
Died:

When they first met in Cambridge, England, in 1951, Francis Crick and James Watson shared a common curiosity: how to fathom the structure of a curious biological molecule called deoxyribonucleic acid (DNA).

Crick had joined the Medical Research Council research group in Cambridge two years earlier, where he began to study the structure of proteins by crystallizing them, passing a beam of X-rays through the crystal, and then analyzing the resulting diffraction pattern that was projected onto a sheet of photographic paper.

By now, scientists were sure that the information about the structure of an organism was stored in a cell's nucleus, passed on from cell to cell as an organism grows, and from parent to offspring in eggs and sperm. There was increasing evidence that the material inside the cell enabling this to occur was DNA, but a fundamental mystery remained: how could any biological molecule store enough information to guide the development of cells, organs, and indeed entire organisms, in a cell that is smaller than a period?

As a sideline to their main projects, Crick and Watson started to consider the structure of DNA. Their interest was enhanced after Watson attended a seminar in London where

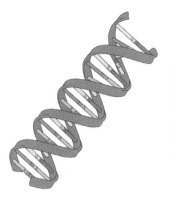

Left: The DNA molecule looks like a spiral ladder where the rungs are formed by base molecules, which occur in pairs. These sequences of base pairs represent the genetic information stored in the DNA.

researcher Rosalind Franklin presented some cross-shaped X-ray diffraction pictures of DNA that indicated that the molecule had a helical structure.

From chemical analysis of DNA, Watson and Crick knew that it consisted of four different components called "bases," and made scale models of each base. They also realized that these bases could form chains, and so they tried building a triple helical arrangement of the components with the spiraling spine formed in the middle, and the rest of the bases pointing out.

The model didn't work, but in 1953 Watson got a sneak-preview of another of Franklin's X-ray diffraction images and realized that the pattern could best be explained by DNA being a *double* helix, with the two strands running in opposing directions.

Crick and Watson jumped to a radical conclusion: the two chains were effectively mirror images of each other. Each chain held equivalent information, and you could produce a second copy by separating the chains and using each as a template. Genetics came of age.

The following fifty years saw a steady increase in the number of techniques that are enabling scientists to make sense of the genetic processes going on inside every living cell. The twenty-first century will see them harness that activity.

Frederick Sanger

Winning one Nobel prize is a rare feat, but to receive two is truly special. Frederick Sanger got his first for showing how amino acids link together to form the protein insulin, and his second for developing a method of sorting out the sequence of molecular "letters" that make up a genetic code.

Born: 1918, Rendcombe, England
Education: Studied Natural Science at the University of Cambridge
Major achievement: Showed how amino acids link to form proteins, and how to read genetic codes
Died:

Born the son of a medical practitioner, Fredrick Sanger initially planned to become a doctor, but before going to university had decided to study science. Arriving at the University of Cambridge he gravitated rapidly to biochemistry and soon joined a team looking at the structure of proteins.

Initially Sanger turned his attention to insulin, the protein involved in diabetes. At the start of his work it was possible to look at the protein using an electron microscope and see its overall shape, or to mash it up chemically into a soup of twenty-two building blocks called "amino acids." Scientists knew that these amino acids were normally linked together in a long chain, but had no clue about the sequence of amino acids within the chain.

Sanger marked the end of the protein chain with a dye that grabbed hold of the last amino acid so strongly that it stuck there even when the chain was dismantled. By studying the amino acid linked to the dye he could say which one was at the beginning of the chain. He then developed a way of breaking the chain into lengths of two, three, four, five, or more amino acids, and identified the amino acid at the end of each fragment. By doing this enough times he determined the order of fifty-one amino acids in a molecule of insulin. The same method could then be used to analyze any other protein of interest.

His second breakthough came when studying another chain-like structure, deoxyribonucleic acid — DNA. This time he wanted to determine the sequence of its "bases," the building blocks from which it is made. The importance of this was that by now, scientists had realized that this sequence spells out the genetic instructions that make up our bodies. To an extent this task was simpler than for proteins, because DNA has only four different bases. But in fact the challenge was greater, because there were many thousands of bases in each sequence.

DNA code is written in four letters, A, C, T, and G. The meaning lies in the sequence of the letters, with most of the information coded in three-letter "words." Most groups of three letters code for an amino acid and some code for where the groups start and stop. For instance, the DNA letters TGC code for the amino acid cysteine, whereas the DNA letters TGG code for amino acid tryptophan. Each of these sequences of three is called a DNA triplet, or codon. The codons that don't code for amino acids provide the grammar of the DNA sequence. For instance, the codon TAA means "stop," essential for telling the cell when to halt producing proteins.

Again Sanger generated fragments of the sequence, this time ending at each different base. By measuring the length of each fragment he showed that you could determine the complete sequence. The process readily lends itself to computer automation, which has allowed scientists to tackle huge sequences, including the three-billion-base sequence that makes up the human genome — the DNA inside the nucleus of each human cell.

Molecule: A collection of two or more atoms held together by chemical bonds.

Amino Acids: The building blocks of proteins; the main material of the body's cells. Insulin is made of 51 amino acids joined together.

The Human Genome Project

From time to time scientists identify a clear goal, but decide that it will be impossible to achieve without a massive joint effort. Sorting out the entire sequence of "bases" in the DNA that makes up human chromosomes was one such undertaking — the Human Genome Project.

Through the course of the twentieth century, scientists began to see how cells could store genetic information by using four different molecular building blocks called bases. They code-named these bases A, C, G, and T, and realized that the secret of the code lay in the sequence in which these bases were ordered. The most common analogy is to think of the bases as letters in a genetic language, and the message is then written in the sequence of these bases.

However, this sequence is vast. In the case of human beings, there are about three billion bases in total. If a letter represented each base, it would take the equivalent of around 6,000 paperback books to write down in full the genetic code of a human being.

The full set of data in a cell is called the "genome." In the 1980s scientists began to devise methods that allowed them to start working out the sequence of bases in patches of the genome. This gave rise to the thought that if enough of them banded together, and if governments chipped in enough money, it would technically be possible to chart the entire set.

After a few years of lobbying, the project finally started in earnest in 1990, with an expectation that it would take around fifteen years to complete. Scientists in eighteen different countries joined in, with the U.S.A. and U.K. taking lead roles. Huge advances in computing power played a major role in chopping that timeline back so that by 1999 the leaders of the project were able to announce that they had the first draft of the sequence. A year later the U.S.A.'s President Bill Clinton and the U.K.'s Prime Minister Tony Blair gave simultaneous press conferences at which they claimed the task was basically complete.

Part of the power of having charted the human genome comes from the fact that these scientists have gone on to determine the sequence of bases in a couple of hundred other organisms. They can then compare the different genomes, looking for similarities and differences, and use this information to study the basic properties of cells that are common to all living things.

With the full human sequence recorded, scientists soon realized that, rather than the anticipated 100,000 genes, there were only 20,000. In itself this discovery showed just how much we need to learn before we can make sense of the ways that cells work. The human genome project has, however, triggered a raft of biotechnology companies each aiming to use genetic knowledge to provide new ways of protecting health and curing disease. As the twenty-first century progresses, this massive genetics project will change the face of medicine.

Edward Jenner

Smallpox was a hugely destructive disease that spread rapidly and killed a third of the people who caught it. Smallpox infection was characterized by fever, aches, and sometimes vomiting. This was followed by a rash, which marked the most contagious period, that progressed to raised bumps and severe blisters. These blisters would scab and fall off after about three weeks, leaving a pitted scar. Many survivors were scarred for life. Physician Edward Jenner found a solution to this feared killer.

Born: 1749, Berkeley, England
Education: St George's Hospital in London
Major achievement: In tackling smallpox, he discovered vaccination
Died: 1823, Berkeley, England

In the eighteenth century no one knew that bacteria or viruses existed, but they did know about smallpox. This disease gave influenza-like symptoms, followed by a rash all over an infected person's body. The rash developed into pus-filled blisters, and the people tended to get infections in the kidneys and lungs before they died.

The only known way of becoming immune was to catch smallpox and survive. In China, doctors deliberately gave people small doses of the disease by grinding smallpox scabs or fleas that had fed on cows with cowpox, and blowing some of the dust into their nostrils. The hope was that they would only have a small bout of illness.

The idea slowly spread west. In the early 1700s English aristocrat Lady Mary Wortley Montague came back from a trip to Turkey having seen women in the Ottoman court making small graze marks on children's arms and wiping the area with smallpox scabs. The children became immune to the disease. Impressed with the idea, she had her five-year-old son treated in Turkey, and on returning to England, she got a surgeon to do the same to her four-year-old daughter.

Although the technique was fairly successful, it wasn't perfect. One in fifty people died as a result of such treatment, and it occasionally triggered smallpox outbreaks.

According to folklore, milkmaids who caught cowpox didn't get smallpox, and on May 14, 1796, Edward Jenner performed a vital, if terribly risky, experiment. He made two half-inch scratches on the arm of an eight-year-old boy and wiped a cowpox scab over the wound. The scab had originally come from the hands of a local milkmaid, Sarah Nelmes. Six weeks later, Jenner exposed the boy to smallpox. He did not become ill.

Virus: Ultra-microscopic infectious agent that replicates itself only within cells of living hosts.

The process worked for both the Chinese and for Jenner because the viruses causing cowpox and smallpox are remarkably similar. When exposed to cowpox viruses the person's immune system developed molecules that are ready and waiting to fight the virus. These same molecules, however, were equally capable of fighting smallpox viruses, so the person developed resistance to both diseases. Vaccination had arrived, and our language still remembers the event: the word "vaccinate" comes from the Latin *vacca*, meaning "cow."

By 1800 some 100,000 people throughout the world had been vaccinated against smallpox, and in the twentieth century the World Health Organization made a concerted effort to wipe out the disease. On October 27, 1977, Ali Maow Maalin, a twenty-three-year-old hospital cook in a small Somali village called Merka, became the last person to catch smallpox. For the rest of the world, smallpox had been consigned to history.

Florence Nightingale

Nursing has not always been thought of as a science, but Florence Nightingale discovered ways to fight disease simply through close observation, the introduction of some basic statistics, and, less simply, her subsequent actions on them.

Born: 1820, Florence, Italy
Education: Taught at home in Cambridge
Major achievement: Polar-area diagrams that record seasonal disease
Died: 1910, London, England

Born into a wealthy, landowning, Unitarian family, Nightingale felt herself called by God to some great cause. Having refused to marry various rich suitors, she eventually persuaded her parents to allow her to become a nurse. In 1851, at age thirty-one, she went to Kaiserwerth, Germany, to study nursing at the Institute of Protestant Deaconesses. Two years later she was appointed resident lady superintendent of a hospital for invalid women in Harley Street, London.

In March 1854 Britain, France, and Turkey declared war on Russia, a conflict which would become the Crimean War. The allies defeated the Russians at the battle of the Alma in September but reports in *The Times* criticized the British medical facilities for the wounded. In response, Sidney Herbert, the secretary at war, appointed Nightingale and thirty-eight other nurses to work at the Barrack Hospital in Scutari, a suburb on the Asian side of Constantinople (the city now known as Istanbul).

Appalled that the wounded men were lying in dirty conditions, often without blankets or decent food, Nightingale complained to the doctors. They took no notice of the upstart nurse, so she began to write to *The Times* in London. The subsequent publicity led to Nightingale being given the task of reorganizing the barracks' hospitals.

At the same time she started recording the number of men who were infected with, and dying from, different diseases.

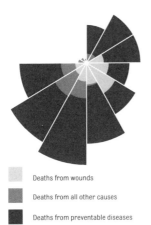

Left: By dividing the year into twelve monthly sectors, Florence Nightingale's revolutionary polar-area diagrams graphically illustrated the huge number of deaths from preventable causes — valuable ammunition in persuading the army to improve hygiene and sanitation on the wards.

Deaths from wounds

Deaths from all other causes

Deaths from preventable diseases

This meant creating systematized processes for collecting data and developing a way of presenting these data visually. Her polar-area diagrams visualized the needless deaths caused by unsanitary conditions. They consisted of a circle divided into twelve sectors, one for each month. Each sector was divided into three zones: the outer area represented the number of deaths from preventable diseases and the inner area showed deaths from wounds. The area of a middle zone indicated death from all other causes. In each monthly segment, the outer region was clearly the largest.

One look at Nightingale's charts and the picture was clear. The numbers dying from secondary diseases exacerbated by poor conditions overshadowed those dying directly from their wounds.

Having data presented clearly enabled her to demonstrate that improving sanitary conditions would decrease the number of deaths. By introducing clean water, fresh fruit, and new systems of hygiene, Nightingale reduced mortality from 60 percent of patients to less than 40 percent. Her legacy is the realization that statistical analysis can provide an organized way of learning and can lead to improvements in healthcare.

Combating Diseases

Louis Pasteur

Each of Pasteur's discoveries represents a link in an uninterrupted chain, beginning with asymmetry in molecules, moving through his realization that some microorganisms cause fermentation in beer and wine — while others lead to diseases — and ending with an understanding of the science behind vaccination.

Born: 1822, Dole, France
Education: École Normale, Paris
Major achievement: Identified the roles played by bacteria in food and disease
Died:1895, Saint-Cloud, France

Like so many geniuses, Pasteur did poorly at school, but excelled once he got to university, and by the age of twenty-six was working on the chemistry and optical properties of asymmetric molecules. These are molecules that chemists say can either occur in left-hand or right-hand form. He came to the conclusion that although many molecules exist in two forms that are mirror images of each other, if they have been produced by a biological process only one of the forms will be present. If both right and left forms are present, a physical process most probably generated the molecules.

This discovery had an unexpected use when he was asked to solve a problem in a factory that was fermenting beets to produce alcohol, because on some occasions the factory instead generated lactic acid. At the time, fermentation was seen as a chemical process that occurred when you brought the right ingredients together: sugar broke down into alcohol as a result of some inherent destabilizing vibrations. The yeast cells found in wine were thought to play no role in the process.

Pasteur realized that the lactic acid crystals he observed were all of one type and correctly concluded that they had been created by living organisms. He then realized that while healthy, round yeast cells generated alcohol, lactic acid was being produced by small, rod-like microorganisms that we now know were bacteria. The solution to the factory owner's problem was to keep everything clean, but the repercussions were that Pasteur realized that microscopic "germs" could have a significant impact on life.

Later, Pasteur was called in to solve problems in the French silkworm industry. The worms were either dying, or failing to spin silk. He found that healthy worms became infected when they nested on leaves used by infected worms. Without fully understanding what was going on he solved the problem by recommending certain conditions of temperature, humidity, ventilation, and quality of food, as well as husbandry techniques that kept newly bred worms away from older ones.

Together this would probably have placed Pasteur in the history books, but it was when he realized that diseases such as cholera, diphtheria, scarlet fever, syphilis, and smallpox were caused by microbiological agents — germs — that his place in history was assured. Remarkably, however, he went further still and showed that injecting heat-damaged bacteria into people made them immune to the diseases caused by healthy bacteria. By doing this he invented modern vaccination and triggered a massive advance in the fight against disease.

Yeast: Single-celled organisms that occur naturally, and are used in baking and brewing industries.

Scarlet fever: A disease that results from infection with a strain of Streptococcus bacteria, spread by respiratory droplets. It carries an erythrogenic (rash-inducing) toxin that causes the skin to shed itself.

Robert Koch

Born the son of a miner, Robert Koch always loved biology, and while studying anthrax he became the first person to demonstrate that specific bacteria caused specific diseases.

Born: 1843, Clausthal, Germany
Education: University of Göttingen
Major achievement: Proved the link between bacteria and disease
Died: 1910, Baden-Baden, Germany

While studying medicine Robert Koch was influenced by Professor of Anatomy Jacob Henle (1809–1885), who believed that living, parasitic organisms caused infectious diseases. After briefly serving as a medical officer with the army in the Franco-Prussian War, Koch became the district medical officer for the province of Wollstein.

Although he was hard-pressed with his medical duties and cut off from other scientific activity, Koch set up a small laboratory in the four-room flat he shared with his wife and young daughter. He started studying anthrax, a disease that was rife among the farm animals in the area. Koch had a hunch that it was associated with a type of bacterium that had recently been discovered.

To start with he took slivers of wood and spiked them into the spleens of animals that had died of anthrax, and then spiked the wood into mice. The mice became infected with bacteria and died. Mice spiked with wood covered with a healthy animal's blood were unaffected. Clearly something in the infected blood was transmitting the disease.

But was it the bacteria or something else that killed the mice? To answer this, Koch developed ways of culturing bacteria and ensured that he could select some that were several generations away from any that had come from an infected animal. These still caused anthrax. The results showed for the first time that it was these bacteria, and nothing else, that caused the disease.

Recognition among the scientific community eventually led to Koch being given proper laboratory space in Berlin. Here he showed that you could grow bacteria on a solid surface such as a potato and on agar kept in a special flat, round dish designed by his colleague Richard Petri (1852–1921) — a container now known throughout the world as the petri dish.

As Koch studied, he came up with a list of four features that a microorganism must satisfy before it can be definitely linked to a disease. Koch's postulates state that you must be able to:

1. *Isolate the organism from every animal that has symptoms of the disease*
2. *Propagate the bacteria in a laboratory*
3. *Reproduce the disease by injecting the organism into a suitable recipient*
4. *Re-isolate the organism from this recipient*

As his work developed, he traveled widely, investigating diseases in various parts of Europe and Africa. When he died, infectious disease had lost some of its mystery and was now the target of scientific research.

Anthrax: Bacterial infection that can cause a severe disease if it enters via the skin and is often fatal if the bacteria or its spores are inhaled.

Agar: Gelatinous material, derived from certain marine algae, that is used as a base on which to grow bacteria.

Disease-Causing Agents

Throughout the history of life on Earth, microorganisms have lived on and inside animals and plants. Sometimes this co-existence benefits both the microbes and the larger organisms. At other times, however, the microbes damage their hosts and cause disease.

For most of human history about half of all people died before they reached adulthood, and most of these deaths were caused by disease. The realization that microorganisms exist, and that these one-celled packages of life were the cause of many terrifying diseases, turned the course of human history. For the first time, societies could start to prevent disease from spreading and researchers could look for therapies that would combat infections.

We now know that many different biological agents can cause disease. The largest such organisms are protozoa — microscopic animals that cause illnesses such as malaria. More common are single-celled bacteria that trigger diseases ranging from irritating sore throats to life-threatening boils. Viruses are even smaller — in fact they are so small that they can only reproduce themselves by hijacking the machinery of living cells. Smaller still are prions, particles of protein which are the infective agent in diseases such as bovine spongiform encephalitis (B.S.E.) and its human equivalent, Creutzfeldt-Jacob disease (C.J.D.).

The first major step on the road to the discovery of infectious organisms was in 1677, when Dutch cloth merchant Antony van Leeuwenhoek (1632–1723) peered through a makeshift microscope at white material he had scraped off his teeth. To his amazement he saw what he called "very little living animalcules, very prettily a-moving." While he was the first person to see a microorganism, no one realized the significance of this observation.

About a hundred years later, English physician Edward Jenner (1749–1823) showed how to combat disease-causing microorganisms by introducing vaccination for smallpox. He used scabs from an infected person to create the treatment, but was unaware that the active ingredient within the scab was a virus.

It took almost another hundred years before French chemist Louis Pasteur (1822–1895) realized that microorganisms were the cause of many diseases. Although physicians increasingly recognized that some diseases such as cholera, diphtheria, scarlet fever, and syphilis were accompanied by the presence of specific microorganisms, it seemed impossible that these tiny cells were the root of the problem. Pasteur, however, had observed microbes causing diseases in silkworms and became convinced that such organisms were also involved in human disease.

While Pasteur was considering the issue, German doctor Robert Koch (1843–1910) discovered that bacteria caused anthrax, a disease that at that point was devastating the European sheep industry. Pasteur went on to show that the anthrax bacteria lived in the earth in some fields and were transferred to the sheep as they grazed. This was the source of the infection.

Through their work, Koch and Pasteur had demonstrated that infectious agents can live in the environment and, if given the chance, can infect animals and humans. At last we had infectious disease in our sights.

Prion: An infectious particle that does not contain DNA or RNA, but consists of only a hydrophobic protein; believed to be the smallest infectious particle known that transmits the disease from one cell to another and from one animal to another.

Alexander Fleming

When Alexander Fleming was born, disease-causing bacteria had terrifying power, not because it was stronger or more lethal, but simply because treatments were unreliable at best. At that time, a simple scratch from a rosebush could be enough to kill you.

Born: 1881, Lochfield, Scotland
Education: St Mary's Hospital Medical School, London
Major achievement: Discovered penicillin, the first known antibiotic
Died: 1955, London, England

After spending four years working in a shipping office, Fleming entered St. Mary's Medical School and trained to be a doctor. He was soon working with vaccine pioneer Almroth Wright (1861–1947) and started looking for substances that could kill bacteria without harming animal tissues.

Hopes of finding chemical cures for diseases had been raised when in 1909 German chemist-physician Paul Ehrlich (1854–1915) had found a chemical that could treat syphilis. He had tried hundreds of compounds, and "salvarsan," the six hundred and sixth, worked. Fleming soon became one of the very few physicians to administer salvarsan in London, and developed such a busy practice he got the nickname "Private 606."

World War I interrupted his work, but after it ended he made the discovery that tears contain lysozyme, a biological molecule that breaks chemical bonds within the cell walls of some bacteria, causing them to burst. It was a significant discovery, but didn't work on all bacteria.

Fleming kept looking and in 1928, while working on the influenza virus, he made a chance observation. Glancing over a set of discarded petri dishes, he noted that an area around a growth of mold was cleared of bacteria. He wondered if it could be the mold that was producing a chemical that killed the bacteria and so took a sample to test. He discovered that this mold was a member of the *Penicillium* family, and that it did indeed release chemicals that, even if they were highly diluted, killed bacteria.

The chemical released turns out to inhibit one particular step in the biochemical process used by many bacteria to build their cell walls. Without this process, the bacteria burst as they try to grow. Penicillin is called a bacteriocidal agent because it actively kills growing bacteria, though it must be noted that it doesn't work on all bacteria, as others use a different biochemical process for building their cell walls.

Having discovered this chemical, Fleming did little more with it, but New Zealander Howard Walter Florey (1898–1968) and an interdisciplinary team that included chemist Ernst Boris Chain (1906–1976) succeeded in extracting penicillin and showed that it had great promise as a treatment.

> A good gulp of hot whiskey at bedtime — it's not very scientific, but it helps.
>
> Alexander Fleming

By now World War II had broken out and with the Battle of Britain in full swing, Florey set up a processing plant first near Oxford, and then in America. Here Florey came across some mold growing on a lump of cantaloupe. This mold produced 3,000 times more penicillin than Fleming's original strain, and American factories were soon making billions of units of penicillin a month.

The discoverer of germ theory, Louis Pasteur, famously said that "fortune favors the prepared mind," a saying which certainly held true for both Fleming and Florey.

Aristarchus of Samos

Although nothing exists of Aristarchus's original workings, we know that he was certainly both a mathematician and astronomer and that he first suggested the heliocentric universe — with planets orbiting the Sun — because of comments made about him and his theories in Archimedes' famous book, *The Sand-Reckoner*.

Born: Ca. 310 BC, Samos, Greece

Education: A student of Strato of Lampsacus, head of Aristotle's Lyceum

Major achievement: The first person to suggest that the planets rotated around the Sun, not the Earth

Died: Ca. 230 BC, Samos, Greece

The commonly held assumption at the time was that the Earth sat at the center of the "universe," and that the whole universe had a radius of the distance between the Earth and the Sun. But Archimedes said that this idea was challenged by Aristarchus, who claimed that the universe was many times bigger, and that the Sun, not the Earth, remained fixed at its center, with the planets orbiting around it.

At that time astronomers believed that all the stars were fixed in position in a single sphere, with the center of the sphere being the Earth. It is as if the ancients thought they were living at the very center of a huge ball that had thousands of tiny holes pricked in the surface allowing light to penetrate from "outside." Aristarchus's idea was new.

His heliocentric idea had power, partly because it was an easy way of explaining why the stars always seemed to be in the same position relative to each other. If the Sun were the center and we on Earth orbited around the Sun, then there should be an apparent shift in the relative positions of the stars. Aristarchus explained this problem away by saying that the distance between Earth and the stars was infinitely large in comparison to the distance between Earth and the Sun. In this case the shift would be so small that you would not be able to see it.

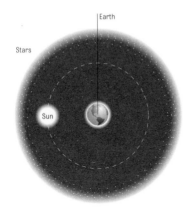

Earth

Stars

Sun

Left: Before Aristarchus proposed his heliocentric (sun-centered) model of the cosmos, astronomers believed that the sun orbited the earth, with a single sphere of fixed stars forming the edge of the universe.

At the time no one took this theory particularly seriously. After all, pointed out Archimedes, this is based on the idea that the universe must be huge, and let's face it, it just can't be that big. Later scientists would realize that it is bigger than even Aristarchus thought.

His only surviving work provides some remarkably detailed geometric arguments that conclude that the Sun was about twenty times farther from the Earth than was the Moon, and twenty times bigger than the Moon. He calculated this by working out the angles in the triangle formed by the Sun, Earth, and Moon, when half of the Moon facing the Earth was illuminated. The Sun is in fact 109 times the size of the Moon and 400 times farther away, but these errors come from the fact that his measuring instruments were poor; his basic calculations were fine. Aristarchus consequently showed that mathematics could help us see our place in the universe.

The Solar System

The Earth is a small but remarkable planet set in the middle of a small and probably unremarkable solar system, tucked into the corner of a medium-sized galaxy which is just one of billions in the universe. Working this out was no mean feat.

When the famous astronomer Nicolaus Copernicus (1474–1543) viewed the heavens he came to a radical conclusion: rather than accepting the ancient view that everything revolved around the Earth, he showed that another explanation for our observations was that the Sun was the central object. When Johannes Kepler produced final proof for Copernicus' theory in 1621, the concept of a solar system became irrefutable.

Even so there was much to discover. At the beginning of the seventeenth century, astronomers had only been able to spot eight bodies that moved across the skies. These were the Sun, Mercury, Venus, Earth and its moon, Mars, Jupiter, and Saturn. Uranus wasn't known until William Herschel spotted it in 1781, Neptune was first seen by Johann Gotfried Galle in 1846, and Pluto by Clyde Tombaugh in 1930.

During this period various observers were beginning to detect moons orbiting many of the planets. In 1610 Galileo spotted Callisto, Europa, Ganymede, and Io all orbiting Jupiter. It was a remarkable feat of observation, even though he managed to miss the other twenty-one bodies that orbit that faraway planet. Indeed each time we get a better view of the solar system we seem to find more objects to name and study. The twin *Voyager* satellites that were launched in 1977 have gradually made their way through the solar system, making close encounters with many of the planets as they travel. Between 1985 and 1989 *Voyager 2* beamed back information about sixteen newly identified major bodies in the solar system, bringing the total known so far to seventy-one.

At a distance of just under six billion kilometers, Pluto is the farthest planet from the Sun. In fact, given its small size and highly elliptical orbit, some people question whether it should really be classified as a planet, but given that it orbits the Sun it is undoubtedly part of the solar system.

As well as beginning to realize the intricacies involved in our own solar system, astronomers have started to see that this is just the start. Each star in the sky represents another sun, many of which we now know are orbited by planets of their own. Our solar system rests in one of the spiraling arms of a medium-sized galaxy that we call the Milky Way. It is about 100,000 light years across and contains some 100 million stars. The Sun orbits around the center of the galaxy, making one rotation every 225 million years.

> The sun, with all those planets revolving around it and dependent on it, can still ripen a bunch of grapes as if it had nothing else in the universe to do.
>
> Galileo Galilei

It turns out that this galaxy is just one of billions of others found within the universe, some of which contain up to three trillion stars. On this scale, our solar system is tiny, for all that it is so important for us.

Claudius Ptolemaeus ("Ptolemy")

While people from different religions and cultures disagreed on many things at the time when Ptolemy was working, they were united on one thing — that the Earth was at the center of the universe.

Born: Ca. AD 100, Alexandria, Egypt
Education: Alexandria, Egypt
Major achievement: Created incredible maps
Died: Ca. AD 170, Alexandria, Egypt

For some this was based on religious belief, for others the conclusion was drawn from philosophical arguments. Ptolemy therefore set out to pull together all previous observations, calculations, and theories, and show how such a geocentric view might work.

When astronomers looked at planets they became aware that they didn't just move smoothly through the sky night after night. Instead there were times when they appeared to stand still for a bit, or even move backward. Ptolemy's explanation was that while the planets orbited the Earth in circles that had centers close to the Earth's center, they also moved in small circles, called epicircles.

To do his work, Ptolemy developed mathematical ways of relating lines and angles, and one of his findings has been handed down as Ptolemy's theorem. This shows that there is a fixed set of relationships in lengths of lines of a four-sided box drawn within a circle. The theory now underpins much current trigonometry.

The system drew heavily on the work of Greek and Babylonian astronomers such as Hipparchus (ca. 190 BC–ca. 120 BC) and required eighty different orbits; the combination of these orbits produced the pattern of behavior seen from Earth. This model of the universe remained unchallenged until the Polish

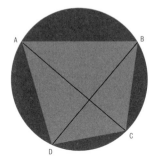

Left: Ptolemy's theorem says that:
(length AD x length BC)
+ (length AB x lengthCD)
= (Length AB x Length BD)
or
AD·BC + AB·CD = AC·BD

scholar Nicolaus Copernicus's (1473–1543) heliocentric view was published in 1543.

In addition to his work in astronomy, Ptolemy was a pioneering map-maker. Convinced that the Earth was a sphere, he developed geometric methods of projecting a sphere onto a flat surface. He also included coordinates of latitude and longitude for every feature drawn on the map, allowing anyone to reproduce it at any scale they wished.

Compared to modern maps, Ptolemy's look distorted, but right through to the sixteenth century his book *Geography* was so far ahead of anything else that it remained the principal work on the subject. His maps cover about a quarter of the world's surface, spanning from the Canary Islands in the west to China in the east, and from the Arctic to equatorial Africa. Indeed *Geography* looked so convincing that it formed a critical part of the thinking that sent Christopher Columbus (1451–1506) sailing west in search of the Indies. Columbus thought that the journey would be short, because on Ptolemy's maps there was only a short span of ocean, but Ptolemy had underestimated the size of the Earth and had overestimated the size of Asia.

Galileo Galilei

To be a free thinker is dangerous in any age. It becomes more problematic when your new ideas challenge established teaching, and when you compound this by acting with a lack of diplomacy.

Born: 1564, Pisa, Italy
Education: Padua University
Major achievement: Realized the importance of taking measurements when developing theories
Died: 1642, Arcetri, Italy

When Galileo Galilei headed off to Padua to study, medical education was still firmly based in the teachings and writings of ancients such as Galen and Aristotle, with little room given for injecting any new ideas. It is hardly surprising then that Galileo soon switched his attentions to the less restrictive world of mathematics.

He turned his mind to studying gravity. To slow down the action of gravity the better to observe it, he ran a ball down a slope rather than simply dropping it. Galileo devised a water clock so that he could accurately measure short time intervals, and found that the time for the ball to travel along the first quarter of the track was the same as that required to complete the remaining three quarters. He realized that the ball was constantly accelerating. By repeating each experiment hundreds of times to see whether the same thing happened each time, he invented an important part of scientific method.

While sitting in Pisa cathedral one day he was distracted by a lantern that was swinging regularly at the end of a chain. He soon found that a pendulum takes the same amount of time to swing from side to side whether you give it a small push or a large push. The frequency only changes if you alter the weight, or change the length of the supporting rope.

Ever the inventor, Galileo developed a two-lens telescope and started studying the Moon, Sun, and planets. He was the first

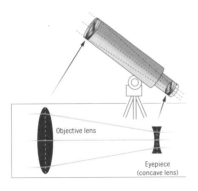

Left: In Galileo's telescope, light rays from a distant object are brought to a focus by a convex lens. A second lens — the eyepiece — then spreads out (magnifies) the focused light so that it covers a larger portion of the viewer's retina, making the observed object appear larger than it actually is.

person to see sunspots, the first to spot the four main satellites of Jupiter, and the first to record that the surface of the Moon had mountains and craters. As he did this he started to look into the work of the Polish-born medic, lawyer, and astronomer Nicolaus Copernicus (1473–1543), who in his 1543 landmark publication said that the planets, including Earth, rotated around the Sun. The Earth was not the center of the universe.

Intrigued by the possibility, Galileo looked for evidence. He noticed that Venus went through phases, much like the phases of the Moon. From this he deduced that Venus must be orbiting the Sun. While he was right in his prediction, he didn't have conclusive proof, and because it contradicted mainstream teaching, the Catholic church instructed him to stop talking about it. Instead he wrote a book presenting his ideas that included a thinly veiled insult toward the pope. The net result was that Galileo was put under house arrest for the rest of his life.

Johannes Kepler

Chiefly remembered for discovering three key laws of planetary motion and making some incredibly accurate astronomic tables, Johannes Kepler also made important discoveries in the fields of optics and mathematics.

Born: 1571, Weil der Stadt, Württemberg (now Germany)
Education: University of Tübingen
Major achievement: Created incredibly precise astronomic tables
Died: 1630, Regensburg (now Germany)

With a father who was a mercenary soldier and a mother who was an innkeeper's daughter, Johannes Kepler's childhood was far from privileged. It was made worse when his father left home to fight in the Netherlands when Kepler was five, and never returned — presumed dead.

Intending to be ordained as a priest, Kepler went on to see his scientific research as a way of fulfilling a Christian duty to understand the works of God, saying at one point that he was merely thinking God's thoughts after him. He was convinced that God had created the universe according to some mathematical plan, and that mathematics would be the route to understanding it.

At Tübingen, Kepler started studying under astronomer Michael Maestlin (1550–1631). To teach some more advanced astronomy Maestlin introduced Nicolaus Copernicus's (1473–1543) recently published idea that the Sun, not the Earth, was the center of the universe. It appears that Kepler instantly liked this theory.

He first developed a complex argument, suggesting that the paths of the planets could be predicted by calculating the sizes of a series of spheres, cubes, and tetrahedrons nested inside each other. The arguments seem curious to twenty-first-century thought, but the results closely correlated with astronomical measurements.

Looking more closely, however, Kepler discovered that Mars's orbit was elliptical rather than circular, with the Sun one of the

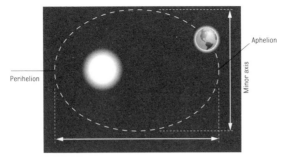

Perihelion

Aphelion

Minor axis

Major axis

Above: Kepler's second law states that the line joining the planet to the Sun sweeps out equal areas in equal times. Hence, the planet moves fastest when it is near perihelion (the point of nearest approach to the Sun) and slowest when it is near aphelion (the furthest point from the Sun).

foci of the ellipse. He produced over a thousand sheets of mathematical workings before reaching this conclusion, and then proceeded to study the other planets. The fact that all the planets' orbits turned out to be ellipses became known as Kepler's first law of planetary motion. His second law was the realization that a line joining a planet to the Sun swept out equal areas in equal times as the planet traveled around its orbit.

Making good astronomical observations requires the best possible telescopes, so there should be little surprise that Kepler also became interested in optics. He discovered ways of making telescopes that used two convex lenses, a design so successful and widely used that it is simply called an astronomical telescope.

While Copernicus had presented the initial idea, and Galileo Galilei (1564–1642) had added some more observational data, it wasn't until Kepler had finished his work that for the first time there was mathematical and scientific proof that the planets orbited the Sun. The universe hadn't changed, but our understanding of it had.

Fredrich Bessel

In January 1799, Fredrich Bessel left school to become an apprentice accountant in an import-export business. Interest in the countries his firm dealt with caused Bessel to study geography, Spanish, and English in the evenings, and he started to ponder how a ship finds its way at sea. It was a short step from these musings to his developing a full-blown interest in astronomy and mathematics.

Born: 1784, Minden, Brandenburg (modern-day Germany)
Education: Left school at 14
Major achievement: Charted 50,000 stars
Died: 1846, Königsberg, Prussia (now Kaliningrad, Russia)

In 1804 Bessel wrote a paper on Halley's comet in which he calculated its orbit using observations made by Thomas Harriot (1560–1621) in 1607. He sent his paper to leading comet expert Heinrich Olbers (1758–1840) who asked Bessel to make further observations and also asked him to consider becoming a professional astronomer. His response was to study celestial mechanics, first at the privately owned Lilienthal Observatory near Bremen and then at the newly built Observatory at Königsberg, where he remained for the rest of his life.

It was in Königsberg that Bessel determined the positions and relative motions of over 50,000 stars. His starting point was the data of English Astronomer Royal James Bradley (1693–1762). This work produced a system of predicting the relative positions of stars and planets. Bessel was one of the first astronomers to realize the importance of working out how many errors were involved in taking measurements. By working out all the sources of error that could be generated by Bradley's instruments, he created a much more accurate set of results. This enabled him to state the positions of stars on particular dates and eliminate from his reckonings such factors as the effects of the Earth's motion.

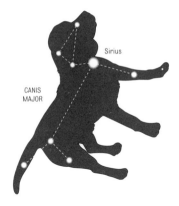

CANIS
MAJOR

Sirius

Left: From the erratic movements of Sirius, a star in the constellation Canis Major ("the great dog"), Bessel deduced the presence of an unseen companion star. Only after Bessel's death did astronomers finally see this "dark star," confirming his prediction.

By eliminating all sources of error — optical, mechanical, and meteorological — Bessel was able to obtain astronomical results of astonishing delicacy from which a great deal of new data could be extracted.

In 1830 Bessel published data showing the positions of thirty-eight stars over the 100-year period from 1750 to 1850. He spotted that two stars, Sirius and Procyon, moved somewhat erratically, and deduced that this variation in their movement must be caused by the tug of previously unseen companion stars orbiting them. He announced in 1841 that Sirius had a companion, and was therefore the first person to predict the existence of so-called dark stars. Ten years later the orbit of the companion star was computed, and astronomers finally managed to see it in 1862, sixteen years after Bessel's death.

Bessel also worked out a method of mathematical analysis involving what is now known as the Bessel function. This helped him to analyze the way that the gravitational forces of three objects interact with each other as they move. The functions have now become an indispensable tool in applied mathematics, physics, and engineering.

Planets and Stars

Edwin Hubble

In the 1920s most of Edwin Hubble's colleagues believed the Milky Way galaxy made up the entire cosmos. But peering deep into space, Hubble realized that the Milky Way is just one of millions of galaxies, and that these galaxies are all rushing away from each other.

Born: 1889, Marshfield, Missouri
Education: University of Chicago and University of Oxford
Major achievement: Showed that the universe is huge and expanding
Died: 1953, San Marino, California

Having studied science in Chicago and Oxford, Edwin Hubble started to examine the stars at Yerkes Observatory in Wisconsin before moving on to the prestigious Mount Wilson Observatory in California, which housed the world's most powerful telescope. The main focus of his attention was on strange, fussy clouds of light called "nebulae."

At Mount Wilson, Hubble found himself working alongside Harlow Shapley (1885–1972), an astronomer who had recently measured the size of the Milky Way. Using bright stars called Cepheid variables as standardized light sources, he had calculated that the galaxy was 300,000 light-years across — ten times bigger than anyone had thought. Shapley was convinced that the Milky Way contained all of the stars and matter in the universe — that there was nothing beyond it. Shapley believed that the luminous nebulae that interested Hubble were just clouds of glowing gas, and they were relatively nearby.

Equipped with his five senses, man explores the universe around him and calls the adventure Science.

Edwin Hubble

In 1924 however, Hubble spotted a Cepheid variable star in the Andromeda nebula, and using Shapley's technique showed that the nebula was nearly a million light-years away — a fact that placed it way outside the Milky Way. We now know that this is the nearest of tens of billions of galaxies.

This alone didn't satisfy Hubble's curiosity. As he studied Andromeda, he realized that the light coming from it was slightly redder than he would have anticipated. The effect is similar to listening to the siren of a moving police car. As it approaches, the tone goes higher, and as it passes the tone drops. A shift toward red is equivalent to a drop in tone. The most likely cause of this so-called red shifting was that the galaxies were moving away from the Milky Way — from our own galaxy. By measuring the shift in all the nebulae he could find, Hubble came to realize that the farther away a galaxy is from Earth, the greater the red shift — in other words, the faster it is moving away from us. The explanation was extraordinary: the entire universe is expanding.

Red shift: The light coming from distant objects is slightly redder than predicted. This red shift is best explained as a lengthening of the wave length of light caused by the universe expanding.

When Einstein heard of Hubble's discovery, he was thrilled. A decade earlier Einstein's new general theory of relativity had predicted that the universe must either be expanding or contracting. Astronomers had told him it was static, so he added an extra "cosmological term" to account for the universe's stability. Hubble had demonstrated that this cosmological term wasn't needed, and that Einstein's own instincts had been right.

(Abbé) Georges Lemaître

Georges Lemaître was originally a priest and his interest in astronomy stemmed from his studies about creation, which, when combined with his scientific and mathematical work, led him to propose the Big Bang theory.

Born: 1894, Charleroi, Belgium
Education: Cambridge University and the Massachusetts Institute of Technology
Major achievement: Introduced the Big Bang theory of the origin of the universe
Died: 1966, Louvain, Belgium

When he started studying the universe, most scientists thought that it was infinite in age and basically unchanged in its general appearance — that it had always been there. Scientists from Isaac Newton (1643–1727) to Albert Einstein (1879–1955) seemed to confirm that the universe had gone on for ever, stable and unchanging.

Lemaître wasn't convinced. Moving to do research at Cambridge, Lemaître reviewed Einstein's general theory of relativity and agreed with Einstein that the universe had to be either shrinking or expanding. Unlike Einstein, who added a cosmological constant to let his equations work in a stable universe, Lemaître decided that the universe was expanding.

He believed this because the idea fitted with early observations of a red shift in color of light from far off galaxies that Lemaître thought could be explained if these galaxies were moving away from us. Lemaître published his calculations and reasoning in 1927, but few people took any notice.

Two years later Edwin Hubble confirmed the existence of red shift and Lemaître then sent a copy of his ideas to Arthur Eddington (1882–1944), a member of the Royal Astronomical Society in London. The British astronomer realized that Lemaître had bridged the gap between observation and theory and the Royal Astronomical Society subsequently published an English translation of Lemaître's paper in 1931. Even so, most scientists agreed with Eddington that the idea of the universe having a beginning was repugnant. Cambridge astronomer Fred Hoyle (1915–2001) coined the term "Big Bang" as a joke.

Lemaître knew that there were still some problems with his theory. To start with he predicted that the universe had expanded at a steady state, but this meant that it was expanding too quickly for the stars and planets to form. Lemaître solved this by using a variation of Einstein's cosmological constant to speed up the expansion of the universe over time. Einstein wasn't impressed. He had dismissed this constant as the worst mistake of his career and was upset to see anyone bringing it back to life. But in 1964 workers at Bell Laboratories in New Jersey became annoyed by a uniform microwave interference that kept being picked up by their radio telescope wherever they pointed it — at one point famously thinking it was caused by pigeon droppings building up in their radio telescope. Eventually, physicist Arno Penzias (1933–) realized that this microwave background interference was a remnant of the Big Bang.

Lemaître heard of this while in hospital recovering from a heart attack. He died two years later, knowing that his theory looked secure.

Big Bang: A theory of cosmology in which the expansion of the universe is presumed to have begun with a primeval explosion.

Einstein's Cosmological Constant: To get a static universe, Einstein added an artificial term, his cosmological constant, to his field equations that stabilized the universe against expansion or contraction.

Arthur C. Clarke

The idea that an object could be fired into space and then constantly fall toward Earth without ever hitting the ground seems a strange one, more fitted to Arthur C. Clarke's famous science fiction, but this concept is the basis behind the geostationary satellites that he brilliantly dreamed into existence.

Born: 1917, Minehead, England
Education: King's College, London
Major achievement: Conceived the idea for satellite communication
Died:

There are times when science fiction and science fact have merged. Often Arthur C. Clarke has been in the middle of the event, either as an instigator or commentator. While he is best known as an author of science-fiction novels, his innovative thinking in the early 1940s led to the prediction that electronic communication would one day use satellites that stood still over the Earth. He wrote about his ideas in an article published in *Wireless World* magazine in 1945.

Although radio and television had been shown to work, Clarke realized that to provide reception across an entire country you would need to have repeater masts positioned every fifty miles, and link these with a massive network of coaxial cables. While that might be financially possible in areas of high population density, there was no way it could be cost effective over whole countries. Supplying a service involving sending programs or messages across oceans was basically impossible. Television was a particular problem — the complex nature of the signal meant that it was more difficult to transmit than radio.

During World War II, Clarke had seen the Germans develop rockets, and these fired his imagination. As he said in his article, if one of these rockets could travel at five miles per second it could "escape the Earth's atmosphere and become an artificial satellite,

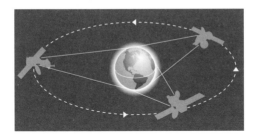

circling the world for ever with no expenditure of power — a second moon, in fact."

If you sent up a rocket containing a communication satellite, and steered it so that it was over the equator, then it could fall in space, and constantly orbit the Earth. Clarke realized that if you positioned it approximately 26,200 miles (42,164 km) from the center of the Earth, in other words approximately 22,237 miles (35,787 km) above mean sea level, then the speed with which it fell would be the same as the speed of the Earth's rotation. Effectively it would hang still above a single point over the equator. By sending enough of these satellites into space, you could generate a network of transmitters that could relay messages to each other and beam them back to Earth with each one serving huge areas.

What started as a suggestion in a small magazine led to the 1965 launch of *Early Bird*, the first commercial geostationary communication satellite, and by 2000 there were over 300 satellites in the so-called Clarke orbit. Electronic communication now spans the globe.

Stephen Hawking

Many people question whether humans will ever understand something as vast as the universe. Bestselling author and astrophysicist Stephen Hawking has gone further than most in developing theories that make sense of current data.

Born: 1942, Oxford, England
Education: University
College, Oxford
Major achievement:
Developed a complex
understanding of the nature
of the universe
Died:

While Stephen Hawking was born in Oxford, most of his academic life has been spent at the University of Cambridge, first in the Institute of Astronomy and then at the Department of Applied Mathematics and Theoretical Physics. He has spent much of his career developing Albert Einstein's theory of general relativity, which had introduced the concept of space-time. In special relativity and general relativity, time and three-dimensional space are treated together as a single four-dimensional concept called space-time. A point in space-time is called an event, and an event must have four reference points: length, breadth, and height, and time.

> Not only does God play dice, but ... he sometimes throws them where they cannot be seen.
>
> Stephen Hawking

Between 1965 and 1970 and working in collaboration with Roger Penrose (1931–), Hawking devised new mathematical techniques to study space-time and then went on to apply this to the study of black holes. These features appear to have been formed by stars collapsing in on themselves and becoming so dense, and their gravitational fields so strong, that, among other things, light can not escape their pull. In 1970 Hawking showed that combining quantum theory and general relativity indicated that black holes can emit radiation. From then on he started working to try to roll quantum theory and general relativity into what people hoped would become a "grand unified theory."

A sign that this may be possible came when Hawking investigated predictions about the creation of the universe that flow from these two theories. He started by calculating that following the Big Bang many objects the size of a proton would be created. While a proton is incredibly small, these particles might have a mass of as much as ten billion tons. The large mass of these mini–black holes would give them huge gravitational attraction and therefore they would be governed by general relativity, but their small size would make them also governed by laws of quantum mechanics.

Hawking's mathematics also show how it is theoretically possible that the universe can be finite, but at the same time have no boundaries or edge. One implication of this is a confirmation that laws of science can be formulated to completely describe the way the universe began.

In 1979 Hawking was appointed Lucasian Professor of Mathematics at Cambridge. The man born on the 300th anniversary of Galileo Galilei's (1564–1642) death now held Isaac Newton's (1643–1727) chair at Cambridge. He rose to fame with the publication, in 1988, of his book *A Brief History of Time*, which became an international bestseller.

Quantum theory: The theory which describes laws of physics that apply on very small scales. The essential feature is that energy, momentum, and angular momentum come in discrete amounts called quanta.

Black hole: An object with such high gravity that not even light can escape. Possibly formed when the most massive of stars die, and their cores collapse into a superdense mass.

Robert Boyle

Having space to research and develop new ideas is costly. You need either a wealthy benefactor or vast private sources of wealth. For Robert Boyle, it was inherited wealth that gave him the freedom to think.

Born: 1627, Lismore, Ireland
Education: Educated at home, in Ireland
Major achievement: Identified the physical nature of gases
Died: 1691, London, England

Born in Lismore Castle in Munster, Ireland, Boyle grew to develop a passion for alchemy, a subject that, in the seventeenth century, was studied by a highly secretive international network of colorful characters who believed that through their studies they would find ways of generating gold from base materials and discover a mechanism or potion that would extend life.

In 1661 Boyle broke from the alchemist's obsession with secrecy and published *The Sceptical Chemist*, a book in which he criticized alchemists for the "experiments whereby vulgar Spagyrists are wont to endeavor to evince their Salt, Sulphur, and Mercury to be the true Principles of Things."

In *The Sceptical Chemist*, Boyle presented the idea that matter consisted of atoms and clusters of atoms. He suggested that these atoms moved around and collided with each other and that these collisions may cause new clusters, with new properties. Crucially he argued that the atoms making up the clusters hadn't changed. Indeed if you got the conditions right you could take these newly created compounds and split them back into their original elements.

While Boyle was writing this book he had also been carrying out experiments with Robert Hooke (1635–1703). The experiments focused on the properties of air and were made possible because Hooke had developed a sophisticated air pump.

Working in Oxford, the pair had placed a lighted candle under a belljar and then pumped out the air. The flame was extinguished as a result. A burning coal in the airless belljar ceased to glow, but reignited if the air was returned before the coal cooled down. Clearly air was needed for items to burn. Using the same equipment, Boyle and Hooke found that air was also important for the transmission of sound. Through an ingenious set of devices, they managed to put a bell inside the jar, pump out the air and then strike the bell. With no air in the jar, the bell was effectively silenced.

Making sense of all of this was difficult, but Boyle did sort out an intriguing relationship between volume and pressure. Boyle realized that to make discoveries you needed to control a situation carefully and change one thing at a time. In studying gas there are four variables that need considering: the amount of gas, its temperature, its pressure, and the volume of the container that is holding it. Boyle fixed the amount of gas and its temperature during his investigations, but varied the volume of the container or the pressure exerted on it. He found that when he halved the container's volume, the pressure of the gas doubled. If he decreased the surrounding pressure, the container's volume increased. We now know this relationship as Boyle's Law — for a fixed mass of gas, pressure and volume are inversely proportional. It formed the basis of all future work on the physical properties of gases.

Alchemy: The ancient predecessor of modern chemistry. The most well-known goal of alchemy was the transmutation of any metal into either gold or silver. Alchemists also tried to create a panacea, a cure for all diseases and a way to prolong life indefinitely. A third goal of many alchemists was the creation of human life.

Isaac Newton

Described in a school report as being idle and inattentive, Isaac Newton grew up to become a formidable mathematician and observer of the physical world. He described universal gravitation, laid the groundwork for classical mechanics, and shares the credit for the development of differential calculus. A genius of the highest order, Newton is regarded as the most influential scientist in history.

Born: 1643, Woolsthorpe, England
Education: University of Cambridge
Major achievement: Did pioneering work on gravitation and mechanics
Died: 1727, London, England

Newton's early schooling may have appeared to be unproductive, but life started to look up when an uncle sent him to grammar school, and his head teacher then pushed him toward Trinity College Cambridge. In 1665 the plague swept across England, and the university closed months after Newton had received his first degree.

Newton returned to Lincolnshire, where myth has it that he saw an apple fall to the ground and started thinking about the force that made it move. After performing a series of experiments he concluded that two bodies, such as an apple and the Earth, or for that matter the moon and the Earth, attract each other with a force that you can calculate by multiplying the masses of the two bodies, and then dividing that figure by the square of the distance between them.

$$\text{Force} = \frac{\text{mass}_1 \times \text{mass}_2}{\text{Distance apart}^2}$$

Squaring values became a frequent aspect of physics equations. To start with Newton did it because it allowed equations to represent his observations. As future scientists began to explain gravitational fields, the idea of squaring started to make physical sense. Gravity can be explained in terms of lines of flux radiating out in every direction from the center of an object; and the more massive an object, the more lines of flux. You can think of these as ropes that grab objects and pull them to the center. At any distance from the object, the gravitational effect of these flux lines will be spread over the imaginary sphere that has its center in the center of the object. The greater the distance from the object, the larger the surface area of the sphere, and the more sparsely spread will be the "ropes."

As the area of the surface of the sphere is defined by the equation Area = $4\pi r^2$ a change that, for example, doubled the distance "r" will quadruple (r^2) the surface area. The gravitation field will then be spread over this much larger area and will consequently be that much weaker.

In 1684 Newton started writing his most famous book, *Mathematical Principles of Natural Philosophy*. In this he extended his ideas and claimed to have identified three of nature's fundamental laws.

1. *That a body at rest, or in uniform motion, will continue in that state unless a force is applied.*
2. *That you can calculate the force applied to an object by measuring the object's weight and the rate at which it accelerates or decelerates.*
3. *That if one body exerts a force on another, the second body will exert an equal and opposite force on the first.*

This Newtonian understanding became the bedrock of physics for three hundred years — and continues to have great value even now.

Physics and Chemistry

Joseph Priestley

The oldest of six children, Joseph Priestley caught tuberculosis as a teenager and had to leave school. He had, however, already learned the basics of Greek, Latin, and Hebrew, and at home taught himself French, Italian, German, Chaldean, Syrian, and Arabic. To this he added basic skills in geometry, algebra, and mathematics.

Born: 1733, Birstall Fieldhead, England
Education: Little formal education
Major achievement: Discovered oxygen
Died: 1804, Northumberland, Pennsylvania

In 1756 he started studying to become a church minister in a liberal section of the Calvinistic branch of Christianity. As an active networker, Priestley thrived on interaction with the great thinkers of the day, and his passion for political thought brought him in contact with Benjamin Franklin in 1766. Franklin had recently flown a kite in a thunderstorm and deduced that lightning was a form of electricity.

Franklin's enthusiasm for this research awakened Priestley's interest in science, and in 1767 Priestley discovered that electricity could pass through graphite. The finding that increased lengths of graphite gave markedly increased resistance to electrical currents led to the creation of millions of resistors that were at the heart of electronics before the invention of the silicon chip.

> In completing one discovery we never fail to get an imperfect knowledge of others... so that we cannot solve one doubt without creating several new ones.
>
> Joseph Priestly

A few years later, the Priestley family moved into a house next to a brewery. The brewing process generates a layer of gas that forms a blanket over the fermenting brew, a gas that was found to extinguish any lighted wood chips that were held in it. Priestley was intrigued and set up a laboratory where he could experiment with the gas. His discovery that it would dissolve in water to produce a pleasantly tangy drink got him noticed in French and English academies of science, and led to the creation of the fizzy drinks industry. The gas was carbon dioxide.

But it was Priestley's work with oxygen that had the most significance. In one experiment, he filled an inverted glass dome with mercury, and then introduced pieces of different materials, from organic matter such as wood to an array of chemicals. Because mercury is so dense, the other materials rose to the top. Priestley then heated the materials by focusing sunlight on them with a magnifying glass. Many materials released a gas that collected above the mercury, and Priestley studied these gases.

In 1774, Priestley collected gas released from mercuric oxide. All other gases he had collected so far extinguished a burning flame, but this one enabled candles to burn more brightly. He called the gas "dephlogisticated air," because at that point people thought that things burned when they lost a hypothetical material called phlogiston. He soon found that plants left in the mercury chamber also produced this gas.

On a tour of Europe he bumped into a fellow examiner of gases, Antoine Lavoisier (1743–1794) and mentioned his work. Lavoisier instantly realized the significance of this gas and named it oxygen. Between them, these two scientists changed the whole course of what we now call chemistry. ·

Algebra: Part of mathematics in which signs and letters are used to represent numbers.

Resistor: A component of an electrical device that offers resistance to electrical current flow.

Alessandro Volta

A world without batteries is almost impossible to imagine, but when Alessandro Volta was born in Italy, no such thing existed. Although Volta didn't speak for the first four years of his life and his family became convinced that he had a mental disability, at the age of twenty-nine he started teaching physics at the local high school, and within months of arriving at the school he had built his first invention.

Born: 1754, Como, Italy
Education: No university education
Major achievement: Created the first electrical battery
Died: 1827, Como, Italy

Named an electrophorus, this device produced an electric charge from friction in a manner similar to the action of rubbing a party balloon on a sweater.

Soon Volta was promoted to professor of physics and three years later moved to a similar position at Pavia University. Here he came into contact with Luigi Galvani (1737–1798), a fellow researcher who had stimulated muscles in the limbs of recently dead animals using electricity. One day, while cutting a frog's leg, Galvani's steel scalpel had touched a brass hook that was holding the leg in place. The leg twitched. Galvani was convinced that this twitch had revealed the effects of what he called "animal electricity" — the life force within the muscles of the frog.

Volta was sceptical and studied whether the electric current could have come from outside the animal. Volta discovered that bringing two different metals together sometimes caused a small electric current to run, and he correctly guessed that this had occurred when Galvani's scalpel touched the hook.

The fact that you could produce electricity without the presence of animal tissue proved that Galvani's idea of animal electricity was wrong, but equally showed that muscles could respond to external stimuli.

Taking the idea further, Volta created a column of alternating silver and zinc discs. He separated the discs with sheets of cardboard soaked in salty water. This stack produced a constantly flowing electric current, and building stacks of varying numbers of elements produced more or less powerful currents. His largest column consisted of sixty layers, but he soon found that having more than twenty elements in the stack produced a current that was painful if you held on to wires attached to either end. What Volta didn't know was that all metals hold on to their electrons with different degrees of tenacity. If you place two different metals next to each other, electrons will flow from the one that is relatively more keen to give them up — this is the start of an electrical current.

Electron: A negative charged particle that orbits the nucleus of an atom.

When Volta demonstrated his stack to the French Academy of Science, the onlookers were so impressed that Napoleon Bonaparte made him the Count of Lombardy. His contribution to the understanding of electricity was so significant that a key measurement of electricity, the volt, was named after him.

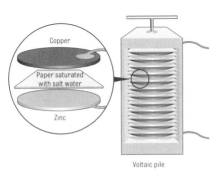

Voltaic pile

Left: The Voltaic pile was the first electric battery. Volta placed together several pairs of copper and zinc discs separated by brine-soaked cloth and noted that the metals and chemicals, in contact with each other, produced an electrical current.

Physics and Chemistry

Electricity

The ability to capture energy in a form that can be fed along wires twenty-four hours a day into homes and places of work has transformed the world and the way we live. Using this electricity to create light removes our dependency on daylight, and employing it in motors and electronic products massively extends the range of tasks that human beings can carry out.

According to Thales, writing in about 600 BC, a form of electricity was known to the Ancient Greeks. They found that rubbing fur on a substance, such as amber, would cause the fur to be attracted to the amber. They also noted that if they rubbed the amber for long enough they could even get a spark to jump. This is the origin of the word "electricity," which stems from the Greek *elektron* (amber).

To understand electricity it is first important to know that all physical matter is made of atoms. They are miniscule, but at the center is a nucleus made up of neutrons and protons, and then flying around this nucleus are smaller particles called electrons. It's difficult to imagine what the atoms are like, but one analogy is that they are like mini solar systems, with the nucleus at the center, and the electrons orbiting, in the same way that planets orbit the sun.

Protons, neutrons, and electrons differ from each other in a number of ways. One is that protons have what scientists call a positive (+) charge, while electrons have a negative (-) charge. Neutrons have no charge, they are neutral. Because the charge of one proton is equal in strength to the charge of one electron, when the number of protons in an atom equals the number of electrons, the atom itself has no overall charge. In this case it is said to be neutral.

Electricity is possible because, while neutrons and protons are held tightly together in the nucleus, one or more of the orbiting electrons can often be removed. In this case the electrons can travel from one atom to another. Materials such as metals are composed of atoms that readily release their electrons, and as a result these electrons can flow through the material easily — the ease of which is described as levels of conductivity. This flow is an electrical current.

Other materials, such as plastics and rubber, are made of atoms that hold on tightly to their electrons. As a result, it is difficult to pass an electrical current through them.

One of the first people to investigate the properties of electricity was Benjamin Franklin (1706–1790). Working with friends in Boston in the 1740s and 1750s he realized that rubbing some materials generated a mysterious force that could produce weird effects such as making a person's hair stand on end. His experiments also merged with a love of playing tricks. He found that you could send a current through water and still use the power to ignite alcohol or gunpowder. He was also known to charge glasses of wine so that the drinkers received shocks.

One day in June 1752, Franklin famously tied a key to the string of a kite and flew it in a thunderstorm. He discovered that he could then gather electrical charge from the key and correctly deduced that a lighting bolt was a vast discharge of electricity. To an extent this was another party trick, but it led the way to realizing that electricity was a powerful part of the natural world, and one that humans would do well to harness.

Michael Faraday

Some people rise to fame by being born into the right family. Michael Faraday did so because he was brilliant and determined. Born in the area of London now called Elephant and Castle, Faraday grew up in a poor but highly religious environment that led him to expect to find a unifying order in the way the world was made.

Born: 1791, Newington, England
Education: No formal education
Major achievement: Pioneered work in electricity and magnetism
Died: 1867, Hampton Court, England

At fourteen, Faraday started work as an apprentice bookbinder, but enjoyed reading the books more than binding them. In one book he found instructions that enabled him to build his own electrostatic machine, and he joined the City Philosophical Society, which met every week to hear and discuss lectures on scientific topics. After attending a Royal Institution lecture given by Humphry Davy (1778–1829), Faraday became Davy's chemical assistant and toured the continent as Davy's valet. Among many scientific luminaries he met on his travels was the aged Volta, who inspired Faraday to investigate electricity when he returned to London in 1815.

In 1820 the Danish natural philosopher Hans Christian Oersted (1777–1851) wrote a paper describing how a compass needle deflects from magnetic north when an electric current is switched on or off in a nearby wire. This showed that electricity passing through a wire generated a magnetic field. In 1821 Faraday took this a step further. He pushed a piece of wire through a cork and floated the cork on water. The ends of the wire made contact with blobs of mercury and through these he was able to transmit electricity to the wire. When a magnet was nearby the wire moved each time he applied a current. Bending

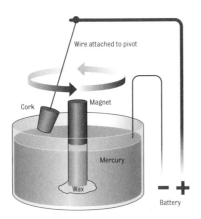

Wire attached to pivot

Cork

Magnet

Mercury

Wax

Battery

– **+**

Left: By passing electricity through a wire in the presence of a magnetic field, Faraday found that a floating cork attached to the wire could be compelled to rotate — a discovery that led to the invention of the electric motor.

the wire, he found a way of making it and the cork rotate when he fed electricity through it and hence discovered electromagnetic rotation, a discovery that led to the invention of electric motors.

Convinced that energy was always conserved within a system, he decided that if electricity could produce a magnetic field, the reverse should also be true — magnetism should be able to produce electricity. It wasn't until almost ten years later that he showed that moving a powerful magnet near to a coil of wire could cause a brief pulse of electricity to flow in the wire — he had discovered electromagnetic induction, the principle behind the electric transformer and generator. This discovery, more than any other, allowed nineteenth-century scientists to turn electricity from a scientific curiosity into a powerful technology.

As Faraday gained fame and reputation he never forgot the excitement of science he had felt as a child, and in 1826 gave the first Royal Institution Christmas Lecture for children — a concept that still carries on to the present day.

James Joule

As the son of a brewer in northern England, Joule had little formal education and never held an academic post. Like so many pioneers of science, Joule set to work with a theological belief that God had built a uniform system — and that it was our duty to work it out and make sense of it.

Born: 1818, Salford, England
Education: Taught by John Dalton
Major achievement: Worked out the relationship between work, heat, and energy
Died: 1889, Sale, England

In 1840 he produced his first major results. By altering the electrical current passing through a wire and the resistance of that wire, he discovered that the amount the wire heated up was proportional to the square of the current; that is to say, doubling the current led to a fourfold increase in heat, while trebling it led to a ninefold increase. He set it out in an equation that is now referred to as Joule's Law.

Energy transferred = the current^2 x the resistance of the wire x the amount of time you ran the current for

The heating is a result of collisions between the moving free electrons and the relatively stationary atoms of the conductor material. Heating increases rapidly as the current increases because the greater rate of flow results in many more collisions.

His expectation of uniform forces in nature led Joule to his next discovery. He was aware that Julius Mayer (1814–1878) had showed that beating paddles in water could cause the water to heat up. Joule set up a paddle that was driven by a falling weight attached to a string. He correctly realized that lifting the weight up stored energy in it, and as the weight fell this energy was transformed by the friction of the beating paddles to the water in the form of heat. In 1849 he read his paper *On the Mechanical*

Space

Atmosphere

Left: Joule realized that friction transforms the kinetic energy of a moving object into a different form of energy: heat. A spectacular example of this is the intense heat caused by the friction of a speeding spacecraft reentering the Earth's atmosphere.

Equivalent of Heat to the Royal Society in London, and was instantly recognized as a name to be reckoned with.

This work contradicted the previously popular caloric theory of heat. This had suggested that heat consisted of caloric, a fluid that could be transferred from one body to another, but could not be created or destroyed.

For seven years, Joule worked with William Thomson (Lord Kelvin) and between them they came to realize that if you allow a gas to expand, it cools. The discovery is known as the Joule-Thomson effect, and it lies at the heart of the mechanism that drives refrigeration systems around the world. The pair also produced a paper that included the first estimation of the speed of gas molecules, a value they set at about 1,500 feet a second for oxygen at average temperatures.

Joule is now immortalized in our use of the term "joule" as a unit of energy, with one joule (1J) being the energy expended when a force of one Newton accelerates an object through one meter: ie 1J=1Nm. In the world of physics, heat and work are now measured in the same units.

Dmitry Mendeleyev

Born in the Siberian town of Tobolsk in 1834, Mendeleyev made his most important contribution to chemistry when he was thirty-five and a teacher at St. Petersburg University. By the time Dmitry Mendeleyev came to do this work, chemistry had moved a long way forward from the days of the seventeenth-century alchemists who believed that materials could be transformed from one to another.

Born: 1834, Tobolsk, Siberia, Russia
Education: University of St. Petersburg
Major achievement: Drew up the first periodic table
Died: 1907, St. Petersburg, Russia

Now chemists were aware that at a chemical level, materials were built of unchanging elements, and these elements combined to make molecules. One question started rising to the top of chemists' minds: why did different materials sometimes look and behave alike?

As a youngster Mendeleyev had moved to St. Petersburg from Siberia after his father died and the family needed to find work. He was the youngest of fourteen children, but worked hard and eventually became a teacher at the University of St. Petersburg and subsequently the professor of chemistry. Here he set about looking at similarities in the behaviors of different elements.

According to Mendeleyev's notes, his periodic table came as a spark of inspiration while he was setting out to write a new chemistry textbook. In a remarkably creative few hours on February 17, 1869, Mendeleyev sat down with sixty-three cards. On each card he wrote the name of one element, its atomic weight, and physical and chemical properties. This pack contained all of the elements known at that time.

By sorting the cards in a grid-like pattern, Mendeleyev realized that you could place them so that the atomic weight increased as you went down a column, but elements in any row shared similar properties. The first column started with lithium, followed by beryllium, boron, carbon, nitrogen, oxygen, and fluorine. In a modern periodic table the grid has been turned sideways and this set of elements appears as the first row. Mendeleyev decided to leave gaps in the table, guessing — correctly — that some elements were yet to be discovered. Equally he grouped some elements according to their properties, even though the weights didn't fit the pattern. He rightly assumed that the recorded weights were wrong. Over time, as the missing elements were found and the weights were recorded accurately, chemists discovered that they matched Mendeleyev's predictions. Designing a system for explaining what you know is clever. Using that explanation to make accurate predictions is remarkable.

Atomic weight: The relative mass of an atom of an element compared to the mass of carbon-12.

It took another half century before the rationale underlying Mendeleyev's observational work gained a foundation in the work of Niels Bohr (1885–1962), whose concept of "orbits" showed that different elements have different numbers of electrons "orbiting" the central nucleus.

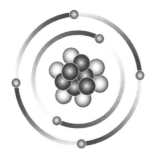

Left: Mendeleyev arranged elements according to their reactive properties and the number of protons in their nucleus. Niels Bohr's subsequent work developed the idea that different elements also have different numbers of electrons "orbiting" the central nucleus, as in this diagram of a carbon atom.

Physics and Chemistry

Wilhelm Röntgen

Some people are prolific problem solvers. Among the problems Wilhelm Röntgen studied were the curious electrical characteristics of quartz, the influence of pressure on the way that fluids refract light, the modification of the planes of polarized light by electromagnetic forces, and the way that oil spreads on water. But he is best remembered for discovering X-rays.

Born: 1845, Lennep, Prussia (now Remscheid, Germany)
Education: University of Zurich
Major achievement: Discovered X-rays
Died: 1923, Munich, Germany

With a Dutch mother and a German father, Röntgen was born in the lower Rhine province of Germany, but moved to Apeldoorn in the Netherlands when he was three years old. He showed no particular skills while at school, but in his spare time enjoyed studying nature and building machines. Not surprisingly, when he had the opportunity of going to university he studied engineering, but soon switched to physics and in time became professor of physics at Würzburg.

In 1895 he was studying what happened when he passed an electric discharge through a chamber containing gas of extremely low pressure. Previous work had shown that doing this with very high voltages could generate a stream of particles that became known as cathode rays. Such chambers have since been refined to form the cathode ray tubes in conventional television screens, and we now call the particles electrons.

On the evening of November 8, 1895, Röntgen enclosed the discharge tube in a thick black carton to exclude all light. When he turned off all the lights in the room, a paper plate coated with barium platinocyanide suddenly became fluorescent. He soon found that the radiation causing this was emitted when cathode rays struck the glass end of the tube, and that these rays had a

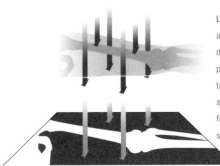

Left: Different body tissues absorb X-ray radiation at different rates. When a photographic plate is exposed to these X-rays, dense objects such as bones that allow fewer rays through to the film show up as white "shadows."

much greater range in air than did the cathode rays. The plate glowed even when it was two meters from the tube.

Intrigued by these rays, Röntgen placed objects of different thickness in their path and found that they exposed photographic plates to varying degrees. Then one day he placed his wife's hand just in front of a photographic plate and shone the rays on it for a short time. When he developed the plate, the result was stunning. There was a ghostly outline of the hand, but the plate also revealed an image of the bones inside her hand and a clear mark caused by a ring she was wearing. It was the first "Röntgenogram" ever taken, and gave doctors an unprecedented ability to look inside people's bodies.

In further experiments, Röntgen went on to discover that the new rays were produced when cathode rays hit a material object. Because the nature of these rays was unknown, he called them X-rays. Later, Max von Laue (1879–1960) and his pupils showed that, like light, X-rays are a form of electromagnetic radiation. The difference is that they have a shorter wavelength than visible light.

Physics and Chemistry

Heinrich Hertz

People who die before the age of thirty-seven don't often leave a huge legacy. Heinrich Hertz was an exception. Born in Hamburg, Germany, Hertz studied at the Universities of Munich and Berlin, and started working with a group of eminent physicists that included Hermann von Helmholtz (1821–1894). While there, Hertz completed a PhD on electromagnetic induction in rotating spheres.

> Born: 1857, Hamburg, Germany
> Education: University of Munich and Berlin
> Major achievement: Made sense of electromagnetic radiation
> Died: 1894, Bonn, Germany

In 1883 Hertz became a lecturer in theoretical physics at the University of Kiel. Here he studied the recent electromagnetic theory of James Clerk Maxwell (1831–1879). This theory was based on unusual mechanical ideas about the "luminiferous ether." This ether was a hypothetical substance supposed to fill all "empty" space, and was thought to be the material that allowed light to travel through the universe.

> One cannot escape the feeling that these mathematical formulas have an independent existence and in intelligence of their own, that they are wiser than we are.
>
> Heinrich Hertz

While another scientist performed some intriguing experiments that proved that this ether didn't exist, Hertz looked at the equations used to make sense of electromagnetic theory. He found that you could reconstruct these so that the idea no longer required ether. Electromagnetic theory had just taken a huge step forward.

Moving to be professor of physics at Karlsruhe University in 1885, Hertz soon discovered the photoelectric effect — where ultraviolet radiation knocks electrons from the surface of metal and creates an electrical current — which is now the basis of many photovoltaic cells used on items from satellites to road signs.

Although Hertz realized the significance of the photoelectric effect, his attention was drawn elsewhere. In 1888, in a corner of his classroom, he generated electric waves using a circuit consisting of a metal rod that had a small gap in it. The gap was small enough for the circuit to be completed by sparks jumping across it. Hertz then showed that these sparks triggered waves of radiation that could be picked up on a second, similar set of apparatus some distance away in the room.

In further experiments he showed that, like light, the waves could be focused or reflected, and that they could pass straight through non-conducting materials. Originally called Hertzian waves, we now think of them as radio waves. Hertz saw no practical use for the discovery, but others were quick to see the relevance. An English mathematical physicist, Oliver Heaviside (1850–1925), said in 1891, "Three years ago, electromagnetic waves were nowhere. Shortly afterward, they were everywhere."

The end result was that a young Italian by the name of Guglielmo Marconi (1874–1937) heard about Hertz's discovery while on holiday in Austria. He rushed home and started developing the idea until he could transmit a signal for more than one mile. In 1901 Marconi transmitted a signal across the Atlantic from Cornwall to Newfoundland, and radio came of age.

Electromagnetic radiation: A propagating wave in space with electric and magnetic components. Electromagnetic radiation is also used as a synonym for electromagnetic waves in general including, for example, light traveling through an optical fiber. Electromagnetic (EM) radiation carries energy and momentum which may be imparted when it interacts with matter.

Marie Curie

At a time when few women had the opportunity to experience the excitement of scientific research, Marie Curie introduced the world to the marvels of radioactivity. Her groundbreaking work sadly led to her death, as she had no way of knowing that radiation emitted from the materials she studied could trigger cancer.

Born: 1867, Warsaw, Poland
Education: The Sorbonne, Paris
Major achievement: Discovered radioactivity
Died: 1934, Sallanchen, France

Born Maria Sklodowska, she grew up in Warsaw but broke free from the oppressive Russian political system — which did not allow women to go to university — by eventually moving to Paris, where she registered her name as Marie on arrival. Here she excelled in physics and math and in 1894 was introduced to the laboratory chief at the Paris Municipal School of Industrial Physics and Chemistry, Pierre Curie, whom she married the following year.

Initially Marie researched the magnetic property of steels. Her focus changed in December 1895 when German physicist Wilhelm Röntgen (1845–1923) discovered X-rays and, almost simultaneously, Frenchman Henri Becquerel (1852–1908) found that minerals containing uranium also gave off unknown rays. While many scientists concentrated on Röntgen's works, Marie studied Becquerel's uranium rays. Setting up a laboratory in a damp storeroom in the Paris Municipal School where Pierre was now a professor, she used a highly sensitive instrument that could measure tiny electrical currents that pass through air which has been bombarded with uranium rays. Pierre and his older brother, Jacques, had invented the device fifteen years earlier. She discovered that the strength of the rays coming from a material depended only on the amount of uranium it contained. In

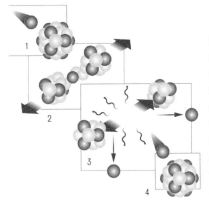

Left: In the reaction shown, a Uranium 235 nucleus is struck by a neutron (1). After absorbing the neutron to become Uranium 236, it splits to form two similar nuclei (2). The split generates huge amounts of radiant energy as well as neutrons (3) that start the process again (4), splitting other nuclei.

addition, the electrical effects of the uranium rays were unaffected if you pulverized the uranium-containing material, kept it pure, reacted it to form a compound, presented it wet or dry, or exposed it to light or heat. Her conclusion was that the ability to give out rays must be a fundamental feature of uranium's atomic structure.

Curie then discovered that other materials gave off rays, and called the phenomenon radioactivity. When she found that a compound called pitchblende gave off more rays than was predicted from the amount of uranium in it, Pierre joined in her research. Together they discovered two new elements in the pitchblende, naming them polonium and radium.

These findings were controversial, but industrial companies started seeing potential in the work. Already they knew that the rays had value in medical imaging, but Marie started to show that they could damage biological tissue, a finding that led to their use in combating cancer. Marie also became aware that radioactive materials were often a source of heat, and started to speculate about the power that was potentially locked up in such substances — energy that other scientists would realize could be released in nuclear power stations and in deadly weapons.

Physics and Chemistry

Ernest Rutherford

When he died suddenly in 1937, the *New York Times* wrote:
"In a generation that witnessed one of the greatest revolutions
in the entire history of science [Rutherford] was universally
acknowledged as the leading explorer of the vast infinitely
complex universe within the atom, a universe
that he was first to penetrate."

Born: 1871, Spring Grove,
New Zealand
Education: Canterbury
College, Christchurch, New
Zealand
Major achievement: Made
sense of atoms
Died: 1937, Cambridge,
England

Having enjoyed a rural upbringing in New Zealand,
Rutherford went on to study and research at several
key universities around the world. But it was while in
his native country that he developed simple but
effective switching mechanisms and monitoring
equipment to determine whether iron was magnetic
at very high frequencies of magnetizing current.

After three failed attempts at getting into
medicine, Rutherford succeeded in picking up a grant to study
science and found himself working with Joseph John Thomson
(1856–1940) in Cambridge University's Cavendish laboratory.
Here Rutherford adapted his detector of "fast transient circuits"
and used it to investigate some of the properties of insulating
materials. Impressed with his ability, Thompson invited him to
join a select team studying the electrical conduction of gases.

During this work, Rutherford discovered that there were two
distinct forms of rays coming from radioactive elements. Passing a
beam of such rays through a magnetic field, he quickly saw that
some were bent, while others traveled straight on. The ones that
went straight on he called alpha particles — which are in fact
helium atoms with their electrons stripped off — while those bent
by the magnetic field he called beta particles, which turned out to
be electrons.

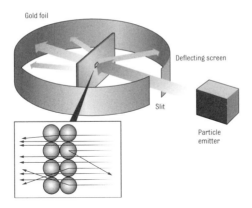

Above: By firing tiny alpha particles at thin films of gold and measuring the resulting deflections, Rutherford deduced that the mass of the gold atoms must be concentrated in an incredibly tiny, dense nucleus.

Moving to McGill University in Montreal, Canada, Rutherford discovered radon, a chemically unreactive but radioactive gas, and published his first book on radioactivity.

It was when he returned to England, this time moving to the University of Manchester, that he had an insight that would change our appreciation of the world. He'd given a student a laboratory practical to run, in which they fired alpha particles at thin films of gold. While most of the particles shot through the gold leaf a few were deflected, while one or two bounced straight back. Rutherford said that this was as if a large naval artillery round had been deflected by a piece of tissue paper.

In 1911 he deduced that this could only have occurred if the mass of the gold atoms was contained in an incredibly tiny nucleus. He left it for a young Danish scientist, Niels Bohr (1885–1962), to add that the rest of the atom would consist of a halo of electrons flying around the nucleus, and he suggested that this was much in the way that planets orbit around a star.

Physics and Chemistry

Nuclear Science

Releasing even a small proportion of the energy stored inside atoms can have an awesome consequence, but it can also be used for peaceful means, as a source of energy.

The realization that there was a class of chemical elements that spontaneously broke down to release energy and radiation gave rise to a new branch of science that grew in importance through the twentieth century. Before this, scientists had worked out that materials were made of atoms, but had concluded that these atoms were unable to change. Now they could see that under certain circumstances some could change.

By 1934 Italian physicist Enrico Fermi (1901–1954) had shown that neutrons could split many different types of atom. But the results he got confused him because he found that adding the mass of the resulting elements gave a result that was much lighter than the original material. In 1938 German scientists Otto Hahn (1879–1968) and Fritz Strassman (1902–1980) produced similar results.

It was a combination of World War II driving all these scientists to work in the United States, together with Einstein's famous $E=mc^2$ equation, that triggered the next development. Einstein's equation showed that it was possible for mass to disappear, so long as energy was given out. At this point Niels Bohr (1885–1962) also arrived in the United States, and joined an active group of physicists who were discussing the possibility of creating a sustainable chain reaction where the energy released by splitting one atom could be used to split others, and so on.

On the morning of December 2, 1942, this group of scientists, led by Fermi, gathered at a squash court beneath the University of Chicago's athletic stadium. In the court they had built a cubic pile of graphite and uranium that had rods of cadmium running through it. The cadmium was included because it absorbed neutrons. They slowly started pulling the cadmium rods out and monitored the temperature of the stack. At 3:25 P.M. the stack temperature rose and they allowed the rods to go back in — they had initiated a self-sustaining nuclear reaction and the world had entered the nuclear age.

While the first uses of this technology were military, with the 1945 bombing of the Japanese cities of Hiroshima and Nagasaki, the scientists were quick to turn their ideas into more peaceful applications. On December 20, 1951, in Idaho, the first power station driven by nuclear energy began to produce electricity, and since then nuclear reactors have been built around the world.

> There are two possible outcomes: if the result confirms the hypothesis, then you've made a measurement. If the result is contrary to the hypothesis, then you've made a discovery.
>
> Enrico Fermi

The debate now rages as to whether we should make more use of nuclear power in producing electricity. It has a major advantage in that it produces no carbon dioxide or other waste gases that could contribute to climate change, but it has the disadvantage of producing small volumes of waste that will continue to emit radiation for thousands of years. It's a prime example of science generating a technology that seems almost too hot to handle.

Albert Einstein

Few people have left such a large mark in the public mindset as Albert Einstein. Born in Germany into a Jewish family, Einstein had little success at school and showed no signs of becoming an international superstar when, in 1901, he took a temporary job as technical expert, third class in the patent office in Bern, Switzerland.

Born: 1879, Ulm,
Württemberg (now
Germany)
Education: University of
Zurich
Major achievement:
Introduced the theories of
special and general
relativity
Died: 1955, Princeton, New
Jersey

His private life at the time was complex. In the same year that he started work in the patent office, his Hungarian girlfriend Mileva became pregnant. Being born outside marriage was taboo, so they gave their child up for adoption, before marrying a couple of years later and then divorcing in 1914. In 1919 he married again, this time to Eva, his cousin, but it appears he was not an easy person to live with. In addition his cultural roots were frequently disturbed. Having started off with German citizenship, Einstein renounced it and became stateless before taking Swiss citizenship. He later reclaimed German citizenship, but the Nazis revoked this because he was a Jew. Finally he moved to America and became a citizen of that nation.

Out of this chaotic life came some stunning scientific insights. In 1905, while working at the patent office, Einstein submitted four papers for publication. His papers on Brownian motion, the photoelectric effect, and special relativity are all probably worthy of winning Nobel prizes, and indeed the Nobel committee did award him the 1921 Prize in Physics for his work on the photoelectric effect. But while the photoelectric effect and Brownian motion had given massive support to those claiming that atoms existed, relativity was something entirely new.

At first he introduced the idea of special relativity, in his 1905 paper "On the Electrodynamics of Moving Bodies." This theory integrated time, distance, mass, and energy and was consistent with electromagnetism, but omitted the force of gravity. It challenged and overturned Newtonian physics by showing how the speed of light was fixed, and was not relative to the movement of the observer. One of the strengths of special relativity is that it can be derived from only two premises:

1. *The speed of light in vacuum is a constant (specifically, 299,792,458 meters per second).*
2. *The laws of physics are the same for all observers in inertial frames.*

Despite its simplicity, it had startling outcomes. Special relativity claimed that there is no such thing as absolute concepts of time and size; observers' appreciation of these features were relative, said Einstein, to their own speed.

In 1915 he took the idea further and developed a theory of general relativity. According to this theory gravity is no longer a force, but a consequence of what he called the curvature of space-time. Unlike special relativity, where reality is different for each observer, general relativity enables all observers to be equal even if they are moving at different speeds.

The ideas are mind-bending, but even though parts of them have been challenged, they formed the grounding for physics throughout the twentieth century.

Theory of special relativity : Einstein's theory that challenged and overturned Newtonian physics by showing how the speed of light was fixed, and was not relative to the movement of the observer.

E=MC²

Ask people to write down a scientific equation and the
chances are they will remember Albert Einstein's formula
E=mc². It is at first glance a simple idea, but one that describes
an incredible concept.

One of Einstein's great insights was that matter and energy are
really different forms of the same thing. This means that matter
can be turned into energy, and energy into matter. The equation
E=mc² grew from experiments that were beginning to show that
light has very strange properties. First, light only exists if it is
moving — there is no such thing as a stationary piece of light.
Secondly, it always travels at a specific speed. Einstein then
claimed that nothing could travel faster than this speed.

Having spotted that the speed of light was a fundamental
value in the universe, Einstein studied the implications of such a
statement. By the turn of the twentieth century, physicists were
comfortable with the idea that the movement energy of an object,
its momentum, could be calculated by multiplying its mass by the
square of its speed. They were used to the idea that in order to
double the speed at which an object traveled, you needed to give it
four times the amount of energy. It was therefore quite natural for
Einstein, when considering the implications of an object moving
at the speed of light, to look at the energy it would contain, and to
apply the same equation. He instantly realized that the resulting
number was huge. After all, the speed of light is 300,000,000
meters per second, so squaring it gives a figure of
90,000,000,000,000,000.

His stroke of genius was to suggest that this meant two
things. First, that if you accelerate an object to close to the speed
of light it will get significantly heavier, and second that you might

be able to get an object to release the energy that is intrinsically present in its mass.

Since he first announced the equation, both concepts have been proved correct. Physicists have built huge particle accelerators that can make tiny protons travel at up to 99.9997 percent of the speed of light. The process demands vast supplies of energy, but at the end, the proton has increased in size by more than 450 times.

The atomic bomb proved the second concept. Physicists saw that in radioactive materials such as uranium, some of the atoms spontaneously fell apart, creating two new elements — but the mass of these new elements was found to be less than the original uranium atom. The difference was released as energy — heat. With World War II raging, it didn't take long for them to see that if you packed enough uranium closely together, you could use this released energy to break up more atoms. This generates a chain reaction that if controlled can drive nuclear power stations — and if uncontrolled leads to an atomic explosion.

> Any intelligent fool can make things bigger and more complex ... It takes a touch of genius — and a lot of courage — to move in the opposite direction.
>
> Albert Einstein

Werner Heisenberg

When Newton looked at forces and movement, he saw predictability, and developed explanations for everyday events. When Einstein reinvestigated the same issues he concluded that reality was more complex, but it could be predicted if you took enough measurements. When Werner Heisenberg helped found quantum theory, he took physics into a world that was much less certain.

Born: 1901, Würzburg, Germany
Education: University of Munich
Major achievement: Developed quantum theory
Died: 1976, Munich, Germany

In the 1920s Germany was an exciting place to work if you loved physics. While still in his early twenties Heisenberg met with the world's biggest names, debating and refining their work. He met Albert Einstein (1879–1955), Niels Bohr (1885–1962), Linus Wolfgang Pauli (1900–1958), and Max Born (1882–1970). Through their inspiration he looked again at the atom. Current theories speculated that electrons orbited a nucleus, much in the way that planets orbit the sun. As he collected data about the way that atoms emit and absorb light he came up with a radically new idea.

Called "quantum mechanics," the new ideas were hotly disputed, but they drew together the mathematics of matrices with the physics of wave mechanics. Attracted to the world center of debate, Heisenberg moved to Copenhagen to join Bohr's group of pioneering physicists. Here he spent a lot of time with Erwin Schrödinger (1887–1961) who visited frequently, and who was also actively trying to make sense of this area of physics.

The more Heisenberg studied the mathematics, the more curious he became. He realized that if you knew the position of an electron, you couldn't say anything about its momentum. Conversely, if you detect an electron's momentum, you won't be

Above: Erwin Schrödinger's famous thought experiment illustrates Heisenberg's uncertainty principle. A cat, a sealed vial of poison, and a small lump of radioactive material are placed in a box. The vial will open if one atom in the radioactive material decays. Since no one can predict when this decay will occur, there is no way of knowing whether the cat is dead or alive. While the box remains closed, the cat is therefore in a "superposition state," being both dead and alive.

able to measure its position. In essence he was claiming that it was always impossible to predict what an electron would do next inside an atom, because of the uncertainty left when you try to measure it.

There are two ways of looking at this. One is to say that the experiments were just not sophisticated enough to do this, but give it a few years and someone would solve the problem. The other was to claim that this was a true reflection of a fundamental property of matter, one he said was described by quantum mechanics. Heisenberg presented this to Wolfgang Pauli in a fourteen-page letter in 1927 and then subsequently presented it to the world. Heisenberg's uncertainty principle had arrived.

Dorothy Crowfoot Hodgkin

Throughout the fields of chemistry and biochemistry, the overall shape of a molecule is known to influence its function, but to know its shape, you need to know its structure — you need to know how the atoms are arranged.

Born: 1910, Cairo, Egypt
Education: University of Oxford
Major achievement: Pioneered X-ray crystallography
Died: 1994, Shipston-on-Stour, England

In the latter part of the nineteenth century, chemists started calculating the shapes of some of the large, carbon-containing compounds that are fundamental to life. They made assumptions about what bonds would form and built scale models to describe the compund structures. Although the exercise worked reasonably well with small molecules, it failed with the larger ones.

In the first years of the twentieth century scientists began to realize that when they shone X-rays through crystals and onto photosensitive paper, they got patterns. The idea is that when X-rays hit a crystal, the electrons surrounding each atom bend the beam. Because there are many atoms arranged in repeating patterns within the crystal, the X-rays produce a series of light and dark patches. Measuring the intensity and relative position of each patch gives clues about the relative positions of atoms within the crystal. Now people could start to make sense of the three-dimensional structure of compounds.

The problem was that some of the most useful — and therefore most valuable — compounds were highly complex, with each molecule containing many hundreds of atoms, each held in a precise location.

Born in Cairo, Egypt, and educated at Oxford University, Crowfoot moved to work with John Desmond Bernal (1901–1971) in Cambridge and started to develop techniques of

X-ray crystallography. One of her early successes was making predictions about the structure of a small protein called pepsin.

After making huge progress in determining how to use X-ray diffraction, Crowfoot moved to Oxford in 1933, and started work on insulin, a task that took her thirty-four years to complete. During this time she married in 1937 and became Dorothy Crowfoot Hodgkin.

In the 1930s and 1940s the germ-busting antibiotic penicillin arrived on the scene. Initial clinical hopes for this drug were dashed because it was difficult to harvest from microbes, and there was no hope of manufacturing it until someone could work out its structure. Hodgkin turned her X-rays on it and found that it had an unusual ring feature, now known as the Beta-lactam structure. This discovery started to give important clues about how the antibiotic worked. In 1955 she took the first X-ray diffraction photos of vitamin B_{12}.

With this pioneering work in X-ray crystallography, Hodgkin was able to discover the chemical structure of penicillin, vitamin B_{12}, and insulin, which enabled them to be manufactured synthetically and become widely available to those in need.

She showed us that we could look inside molecules and see their structure. From here it is a short step to predict their function and work out how to design chemical drugs that could affect them. Biochemistry and the pharmaceutical industry would never be the same again.

Pythagoras

Nothing remains of Pythagoras's own work, but early biographies show that as a child Pythagoras of Samos traveled widely with his father. He learned to recite Homer and spent time with some of the key thinkers of the day.

Born: Ca. 580 BC, Samos, Ionia, Greece
Education: Picked it up as he traveled
Major achievement: Turned counting into mathematics and shapes into geometry
Died: Ca. 500 BC, Metapontium, Magna Graecia (now Italy)

In his mid-thirties he visited Egypt after the tyrant Polycrates seized control of the city of Samos. There he developed his interests in mathematics and religion before setting up his own society on the east of the heel of southern Italy. This society had an inner circle of followers known as *mathematikoi* who held a set of core beliefs. Among them was the idea that, at its deepest level, reality is mathematical in nature.

Pythagoras was passionate about principles of mathematics and the concept of "number." He made the remarkable step of moving from counting objects — two sheep plus two more sheep equals four sheep — to saying there is a universal principal that 2 + 2 = 4. The numbers 2 and 4 became abstract quantities, and Pythagoras has gone down in history as the first person to realize their importance and power.

He studied musical instruments and noticed that strings produce notes in harmony with each other when the ratios of the lengths of the string are whole numbers. The same thing happens with wind instruments, where you get harmonies when the lengths of the tubes are whole-number ratios of each other. These sorts of findings led him to conclude that everything in the cosmos was in fact a number.

According to Pythagoras, numbers had personalities. Some were masculine or feminine, perfect or incomplete, beautiful or ugly. Ten was the very best number. To start with, it contained in itself the first four integers — one, two, three, and four: 1 + 2 + 3 + 4 = 10. If you wrote these down as a series of dots it formed a perfect triangle, one that looked the same from any side.

One of his greatest contributions, however, was his famous geometry theorem, a theory that drew together many aspects of ancient teaching. This states that if you make three squares where the length of the sides are the same as the lengths of the three sides of a right-angled triangle, the area of the square relating to the longest line (the hypotenuse) is equal to the two other squares added together. It's a simple idea, but as the realization of the rigid strength of triangles has grown over the centuries, it has become more and more important in math, design, and construction. The triangles drawn in ancient sand still have relevance today.

> **Geometry:** The pure mathematics of points, lines, curves, and surfaces.

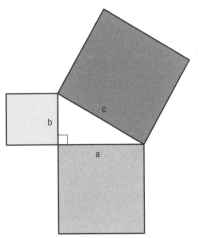

Left: Pythagoras's Theorem states that the sum of the areas of two squares that make up two sides of a right-angle triangle equals the area of the square on the side opposite the right-angle (hypotenuse). In algebraic terms, $a^2 + b^2 = c^2$ where c is the hypotenuse. The theorem is of fundamental importance in Euclidean Geometry as a basis for the definition of distance between two points.

Mathematics

Euclid of Alexandria

Very little is known about Euclid as a person, and almost anything that is said of him is based more on speculation than certainty. History remembers him chiefly because of his treatise *The Elements*, in which he drew together all preceding mathematical thought in such a thorough way that it became a key text for the next 2,000 years.

Born: Ca. 325 BC
Education: Alexandria
Major achievement:
Collected together all
previous mathematic
knowledge
Died: Ca. 265 BC, Alexandria,
Egypt

In fact, Euclid probably wrote *The Elements* by copying parts of several different textbooks. In it he takes for granted that points, lines, and circles exist and makes no attempt to justify or define them, concentrating instead on defining geometric shapes such as an oblong, a parallelogram, and a rhombus.

However, the part of his book that has kept mathematicians busy for centuries is his "five postulates." The first three recognize some features of lines that may seem obvious to us now, but part of the role of mathematics is to make sure that the obvious is in fact true. Euclid's postulates state that:

1. *You can always draw a straight line to join any two points.*
2. *You can always extend a straight line indefinitely in a straight line.*
3. *For any straight line, you can draw a circle that uses it as a radius, with one endpoint at the center and the other on the circle itself.*

Postulates four and five are subtly different from the first three because they start to make assumptions about the nature of three-dimensional space.

4. All right angles are the same as each other.

The importance of this is the recognition that wherever you put a right angle it will be the same, which means that the space it exists in must be the same all over the universe.

5. You can draw one, but only one, line through a point if that line has to run parallel to another line.

This so-called fifth postulate has caused a huge amount of debate over the centuries because mathematically it cannot be proven. Indeed in 1823 a Hungarian mathematician, Janos Bolyai (1802–1860), showed that in certain circumstances the postulate did not hold. It has consequently been dropped from the official list of postulates.

Together the first four postulates form the foundation of all geometry, a subject that has grown by establishing a small number of accepted truths, and then building on them in a step-by-step process.

The result is far more than mathematical intrigue, but a deeper understanding of the structure of the natural world, and an ability to experiment with new ideas in design and construction. Euclid may not have been an outstanding mathematician, but his writing makes him the most influential teacher of math in history.

Rhombus: A plane four-sided figure in which all sides are the same length, and opposite sides are parallel.

Parallelogram: A plane four-sided figure in which the opposite sides are parallel to each other and of equal length.

Archimedes of Syracuse

Some people get hooked on theory, others on application, but Archimedes proved that he was a supreme mathematician as well as an exceptional engineer. His mathematical ideas are still employed today, as are some of his inventions, such as those for moving water and lifting weights.

Born: Ca. 290 BC, Syracuse, Sicily
Education: Alexandria, Egypt
Major achievement: Linked mathematics to inventions
Died: Ca. 212 BC, Syracuse, Sicily

Ironically we know most about Archimedes from biographers of the Roman soldier Marcellus, who was in charge of the army that killed him. It appears that Marcellus was impressed with Archimedes because his inventions and war machines had enabled the people of Syracuse, in Sicily, to put up such a fight against him. During the Roman siege of Syracuse, Archimedes is said to have built arrays of hinged mirrors that enabled him to focus sunlight onto the attacking ships. Like using a magnifying glass to focus sunlight on a stick on a summer afternoon, the wood got hot and started to smoulder. Apparently Archimedes' mirrors were so effective that the ships caught fire.

If any of the ships escaped his focused light, then they were caught by another of his inventions. This was some form of lever system reaching out over the sea, which grabbed the side of a boat and rolled it over. Unfortunately there is no record of the exact design of this lever, and this has been the source of much speculation.

His invention for moving water was less aggressive in nature. Archimedes was fascinated by the mathematics of spirals. He realized that if you confined a spiral in a tube, laid it at a slight angle with the lower end in some water, and turned it, then the water could be drawn up the tube. Now known as an

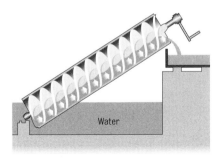

"Archimedes screw," this system is still used throughout the world as a simple method of moving water and commodities such as grain.

It was mathematics that really excited Archimedes and he performed numerous geometric proofs using Euclid's (325 BC–265 BC) ideas. He was especially proud of his discovery that the volume of a sphere is two-thirds the volume of the smallest cylinder that can be placed around it.

According to legend, Archimedes discovered his principle of buoyancy while taking a bath. This principle states that objects of equal size displace the same amount of water, and that something floating in water will displace its own weight of water. It has become central to all boat design ever since.

Archimedes was obsessive about his work. Most commentators agree that he died when Roman soldiers burst into his house, having overrun Syracuse. They had been ordered not to kill Archimedes, and so asked his name. Instead of answering, Archimedes attempted to protect the writing he was working on in the sand on the floor, saying, "don't disturb my circles." In an act of impatient violence, a soldier struck him down.

Pi

Pi (π) is an enigmatic number. It will never be recorded precisely, but allows mathematicians to make sense of equations involving circles and spheres. Its power extends through many branches of mathematics as well as reaching into applied sciences such as engineering and architecture.

This one number relates the length of the distance around a circle, its circumference, to the distance across the circle — its diameter. To mathematicians and engineers it is lovingly knows as pi. The beauty of pi is that it allows you to calculate the circumference of a circle if you know its diameter, or vice versa.

Circumference = pi x diameter or C=πd

Or its area would be:

Area of the circle = $\pi(\frac{d}{2})^2$

You can take this a step further and use pi to calculate the volume of a sphere, so long as you know the sphere's diameter.

Volume = $\frac{4}{3}\pi(\frac{d}{2})^3$

Recognition of this number stretches back at least as far as the Babylonians in 2000 BC, where its use is recorded in the Rhind papyrus. This document is covered in mathematical calculations and was found by Egyptologist Alexander Henry Rhind, who bought it in Egypt in 1858.

The first person to calculate the ratio and give pi its numbers, however, was Archimedes (287 BC–212 BC). He became aware that this number was going to be hard to track down and ended up concluding that it was greater than $^{223}/_{71}$ (ie 3.1408), but less than $^{22}/_{7}$ (i.e. 3.1428).

Archimedes arrived at this by drawing pairs of polygons where one fits inside a circle and the other fits on the outside. As a rough approximation, you could start by using a square, a four-sided polygon. The larger square has sides that are the same length as the diameter of the circle, while the smaller square has a diagonal that is the same length as the circle's diameter. This allowed him to estimate the ratio between the circle's diameter and its circumference. The more sides used in the polygon, the closer the two values, and the better the value of pi.

The number has a practical use, but it has also developed a life of its own, because of the difficulty of defining it exactly. In the sixteenth century Machin calculated the value of pi to 100 decimal places and von Vega extended it to 140. Now that computers have joined in the game, mathematicians have tracked down the value of pi to a staggering 206,158,430,000 decimal places.

One intriguing feature of pi is that there is no pattern or repeating sequence in these numbers. Consequently mathematicians have played various games that enable them to remember the number. Some involve mnemonics such as *How I want a drink, alcoholic of course, after the heavy lectures involving quantum mechanics*, which, if you count the number of letters in each word, enables you to recall the number 3.14159265358979. Alternatively some people just memorize the string of numbers. The world record is held by Japanese mathematician Jirojuki Goto, who in 1989 took ten hours and fifty minutes to recite pi to 42,195 decimal places.

Mathematics

Hypatia

The years that preceded the rise of Christianity and the demise of the Roman Empire were packed with passionate debate. Ideas and beliefs were held with fervor, and defended vigorously. For many key leaders the most important thing in life was to be physically healthy and intellectually powerful.

Born: Ca. 370, Alexandria, Egypt
Education: Alexandria
Major achievement: Developed astrolabes, and made key observations about the planes of regular solid shapes
Died: Ca. 415, Alexandria, Egypt

It was into this situation that Hypatia was born. Like so many aspects of her life, the date of her birth is hotly disputed, with estimates ranging from AD 355 to AD 370. Most agree that Theon, her father, keen to raise the perfect human being, schooled her in mathematics and philosophy and developed a physical training program to ensure a healthy body.

While most scientists make history because of their work or the way they lived, Hypatia is best known for the way she died — her involvement with philosophy and mathematics that brought about her violent end at the hands of religious zealots. According to an ancient text called the Suda lexicon, Hypatia wrote commentaries on existing mathematical books, and concentrated a lot of her work on the mathematics of cones. She was interested in studying the planes that were formed when she sliced through a cone, and extended this to look at hyperbolas, parabolas, and ellipses.

Like most philosophers of the time, she looked to the sky and studied Ptolemy's work. While most of her own writings no longer exist, we can still get a flavor of some of her work from letters written by Syrenius, one of her students. These letters claim that she spent a lot of time developing astrolabes, mechanical devices that could show how the sky appears at a specific place at a given time. The astrolabe consists of a drawing of the sky marked in such a way that positions in the sky are easy to find. By adjusting the movable components to a specific date and time, you get a representation of the way the stars and planets will be aligned. Astrolabes therefore enabled people to solve astronomical problems in a visual way.

Politics and philosophy are sometimes difficult to separate, and Hypatia's association with the civil governor Orestes angered some people in the Christian community. Rumors and counter-rumors spread until one day a mob attacked Hypatia on her way home and dragged her to a church, where she was brutally murdered and dismembered, and her remains burned without ceremony.

Her death and the loss of an important library of scrolls, which occurred around this time, mark a low point in the history of philosophy and science. It would be centuries before another woman had the opportunity to rise to the top of the scientific tree.

Hyperbola: An open curve in which all points have a constant difference in distance from two fixed points called focuses.

Parabola: A type of curve, any point of which is equally distant from a fixed point, called the focus, and a fixed straight line, called the directrix.

Below: Building on the work of Apollonius, Hypatia demonstrated that all of the common curves can be created by slicing through a cone, known as conic sectioning.

Parabola   Circle   Hyperbola

George Boole

Self-taught English mathematician George Boole moved logic from the realm of philosophy and introduced it to mathematics. His so-called Boolean algebra now forms the basis of systems ranging from electronic circuits to Internet search engines.

Born: 1815, Lincoln, England
Education: Self-taught
Major achievement:
Introduced Boolean logic
Died: 1864, Ballintemple,
Ireland

Born in Lincoln, England, George Boole had little formal education, but started to teach himself mathematics at the age of ten and was soon studying the works of Isaac Newton (1642–1727) as well as French mathematicians Pierre Simon Laplace (1749–1827) and Joseph Louis de Lagrange (1736–1813). However, he knew little of the systems of logical debate that had been developed by the likes of Plato and Socrates, so Boole's ideas were not restricted by previous patterns of thought.

As he puzzled over ideas, Boole argued that there was a close analogy between algebraic symbols and symbols that represent logical interactions. He developed a very simple process of analysis that allowed people to break thought processes into individual small steps. Each step involves deciding whether a statement is either true or false. If true, an answer could be assigned a value of 1 and if false, 0.

The power of the technique is that you can then add answers simply by using one of three operators, AND, OR, or NOT. The AND operator will return a value of 1 if both of two statements are true, while the OR operator will return a value of 1 if only one of the terms is correct. Conversely the NOT operator will return a value of 1 only if one is true but the other is false. By adding many steps, Boolean algebra can form complex decision trees that produce logical outcomes from previously unrelated inputs.

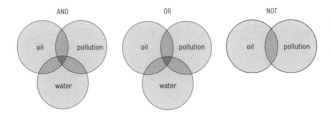

AND	OR	NOT
oil · pollution · water	oil · pollution · water	oil · pollution

Above: In Boolean logic, AND is used to locate instances where all terms are present, so would retrieve records that discuss all three terms oil, water, and pollution. OR is used to locate instances where any one of the terms is present so would retrieve documents that discuss one or more of oil, water, or pollution. NOT is used to locate instances where the first term is present but the second is omitted, so would retrieve documents that discuss oil but not pollution.

In 1847, and while still working as a school teacher in Lincoln, Boole published *Mathematical Analysis of Logic*. This small book presented his idea that logic was better handled by mathematics than metaphysics. The book got him noticed, and even though he didn't have a degree he was offered the post of professor of mathematics at Queen's College, Cork. There he developed his ideas and published his masterpiece *An Investigation of the Laws of Thought*.

One day Boole walked home in drenching rain and caught pneumonia. His wife, who was the niece of the pioneering explorer Sir George Everest, believed that the best way to treat an illness was to expose the person to the original cause. She therefore frequently poured water over his bedclothes to ensure they remained wet. Unsurprisingly, he died.

While he was excited about his work, Boole would have been amazed to see the incredible practical use that has been made of his logic.

Jules-Henri Poincaré

Poincaré started university in Paris studying mining engineering, mathematics, and physics. Moving up the academic ladder, he later held the position of professor in physical and experimental mechanics, mathematical physics and theory of probability, and celestial mechanics and astronomy. He was one of those rare thinkers able to make major contributions to fields as diverse as analysis, algebra, topology, astronomy, and theoretical physics.

Born: 1854, Nancy, France
Education: École
Polytechnique, Paris
Major achievement:
Pioneered the mathematics
of chaos
Died: 1912, Paris, France

As part of his work, Poincaré developed preliminary versions of the theories of relativity and special relativity that were made popular by Albert Einstein. He was also alive long enough to see the arrival of quantum theory and was ideally qualified to grasp that this new way of understanding matter had turned Isaac Newton's simple mechanistic systems upside down.

At the beginning of his scientific career, Poincaré devised a new way of studying the properties of functions defined by differential equations, and was the first person to study the general geometric properties of these functions. He went on to apply this thinking to his study of the stability of the solar system — a piece of work that was stimulated by a competition established by the king of Sweden.

One of the king's questions asked contestants to show that Newton's equations describing the solar system would enable planets to move around, but overall to be stable. The question effectively asked people to prove one of the most difficult problems in mathematical physics, the famous *three body problem*. Solving it meant drawing together nine simultaneous differential equations. While Poincaré failed to give a complete solution, his work was so impressive the judges gave him the prize anyway, because they thought that his work would establish a new era in the history of celestial mechanics — the study of the motion and gravitational interaction of bodies in space, used to calculate the orbits of objects in space and predict their motion.

This research on the stability of the solar system stimulated Poincaré to look at chaos within other systems, but his notion of chaos is far more complex and controlled than simply pointing to a breakdown in order. Poincaré believed that if we knew all the laws of nature and the exact state of the universe at its very first moment, we would be able to predict the situation in the universe at any other moment. But, he said, even if we knew the laws that well we would still have a problem, because a tiny misreading of the initial situation could result in a huge miscalculation of the subsequent outcome. The consequence of this, as any weather forecaster is keen to emphasize, is that it is always very difficult to make accurate predictions.

During his career Poincaré became one of France's most accomplished theoretical scientists, and turned chaos and unpredictability into a subject of mathematical study.

Topology: The physical layout of a network, that is, the way in which constituent parts are related or arranged.

Chaos theory: A mathematical theory in which a high level of order and pattern is mapped onto chaotic systems. The theory has been applied to weather forecasting, business cycles, fluid motion, planetary orbits, electrical currents, medical conditions, and even the arms race.

Andrey Nikolayevich Kolmogorov

Reconciling the precision of mathematics with apparently random processes seems a hopeless task, but Andrey Kolmogorov did just that, and formed the mathematical basis of probability theory.

Born: 1903, Tambov, Russia
Education: Moscow University
Major achievement: Created probability theory
Died: 1987, Moscow, Russia

After his mother died giving birth to him, Andrey Kolmogorov was brought up by his maternal aunt in the village of Tunoshna, on the river Volga. He had a standard schooling, but within a couple of years of starting to study at the University of Moscow, Kolmogorov had shown that he was going to be no ordinary student. By the age of nineteen he was already becoming an international celebrity in the world of mathematics. On leaving postgraduate school in 1929 he had already published eighteen academic papers on a wide variety of areas of mathematical study, each of which was regarded with awe by specialists in the particular field.

> The theory of probability as mathematical discipline can and should be developed from axioms in exactly the same way as Geometry and Algebra.
>
> Andrey Kolmogorov

While on a three-week boating trip down the Volga in 1929, Kolmogorov started working on a new way of looking at the apparently random movement of molecules in gas that is known as Brownian motion. This phenomenon had first been described by Robert Brown (1773–1858) in 1827, and various notable people had puzzled over it, including Bertrand Russell (1872–1970) and Albert Einstein (1879–1955).

To make sense of Brownian motion, Kolmogorov developed a new set of theories based around traditional ideas of probability. This led him to lay the mathematical foundations for modern probability theory and the study of stochastic processes, where the change in one random variable influences all other subsequent events.

Brownian motion: The random motion of microscopic particles suspended in a liquid or gas, caused by collision with molecules of the surrounding medium.

His aim was to create a watertight system of self-supporting mathematical arguments that made sense of apparently random events. His work started to spread out of the realm of theoretical mathematics and helped geneticists to predict the way that characteristics were inherited in plants and animals. Inheritance was realized to be governed by complex rules of probability and Kolmogorov's work gave a framework for this biological thinking. He showed the mathematical basis for the laws of inheritance that Gregor Mendel (1822–1884) had deduced from his experiments on peas and flowers.

Kolmogorov was passionate about his work and keen to help educate others. By the time he died he had taught nearly seventy research students. His work may be buried in deep mathematical theory, but its applications have influenced a wide variety of disciplines. He went on to apply it to areas of information theory, as well as looking at probability in linguistics, prose- and verse-style, and speech. Not many people may have heard of him, but few are untouched by his work.

John von Neumann

Being of Jewish decent and living in central Europe in the early part of the twentieth century meant that the von Neumann family found themselves moving between Hungary, Austria, Germany, and Switzerland to avoid politically motivated conflict.

Having initially been named János and called Jancsi as a child, he changed his name to John when, at age twenty-six, he arrived in America in 1930 to start work at Princeton.

Born: 1903, Budapest, Hungary
Education: University of Berlin
Major achievement: Developed game theory
Died: 1957, Washington D.C.

While he contributed to many branches of math, it was his invention of "game theory" that stands as his chief achievement. Working with Austrian-born Oskar Morgenstern (1902–1976), von Neumann looked at the way people negotiate and interact with each other, and developed mathematical models that described what he saw. The term "games" has now become a scientific metaphor for a wide range of human interactions in which the outcomes of an encounter depend on the interactive strategies of two or more people who have opposed, or at best mixed, motives.

His initial area of interest was economics, where previous work looking at how people behaved was based on the assumption that human beings are absolutely rational in the way they make choices. The theory was that each person would do all they could to maximize his or her profits, incomes, or enjoyment; an individual needs to consider only his or her own situation and the "conditions of the market." The problem was that the theory didn't work in cases where competition in the market was restricted or the rights of individuals were poorly defined.

Game theory confronted this problem and provided an explanation of economic and strategic behavior when people interact directly, rather than "through the market." To do this, von Neumann looked at situations where people (agents) made strategic decisions based on the actions of other people (live variables). This situation was very different to the previous thinking that had assumed that agents made decisions by looking at fixed conditions, such as the price of an item (dead variables).

All of von Neumann's work in game theory concentrated on what is now called non-cooperative or "strategic" games. These are situations where the various people involved are actively seeking to do better than each other. Later, mathematician and Nobel Prize winner John Nash (1928–) developed a second area that looked at cooperative games, where each person involved is seeking to do what is best for all.

The power of game theory is shown, for example, in the fact that it is now applied by people studying the way investors work within a stock market, armies operate in a battlefield, or individuals work within families and societies. The range of applications has proved far beyond the intitial expectations.

> If people do not believe that mathematics is simple, it is only because they do not realize how complicated life is.
>
> John von Neumann

Alan Turing

Following a familiar scholastic pattern for great thinkers, Alan Turing was bored by schooling, had scruffy handwriting, and struggled in English classes. In mathematics and science he preferred chasing his own ideas rather than solving the simplistic problems set by the teachers. This desire to find his own solutions proved useful as he ended up devising a hypothetical machine to solve a particular mathematical problem. The idea was revolutionary, and this machine turned into what we now call a computer.

Born: 1912 in London, England
Education: Studied mathematics at the University of Cambridge
Major achievement: Dreamed up the universal calculating machine
Died: 1954, Wilmslow, England

Two years into his course in mathematics at the University of Cambridge, Turing learned about a question posed by David Hilbert (1862–1943). It was the question of Decidability — the famous *Entscheidungsproblem*: is it possible to find a definite method for deciding whether any given mathematical assertion is provable?

To start with, Turing analyzed how people performed methodical processes and realized that generally people acted mechanically. He then jumped to imagine a theoretical machine that could perform certain precisely defined, simple tasks by reading and writing symbols on paper tape. This, he said, would be able to do everything that would count as a "definite method," and his definite method has become what we would now call an algorithm.

Even though the world was decades away from seeing its first computer, Turing conceived the idea of a universal machine that, when loaded with a set of coded instructions, would solve a particular problem. It was universal because loading it with a different set of instructions would get it to perform a different task. Turing's idea created the intellectual framework required to write a computer program. Within a decade technology had started to provide solutions in the form of primitive computing devices.

One of the first machines consisted of a set of electromagnetic relays that could multiply binary numbers, and during World War II the British government got him to produce a mechanical means of making logical deductions for breaking codes that the Germans were constantly inventing. From late 1940 onward he and colleagues had developed their "Bombe," a machine which made it easier to read Luftwaffe signals. By the end of the war, Turing had overseen the development of the first digital electronic machines, in the form of the Colossus project.

> A computer would deserve to be called intelligent if it could deceive a human into believing that it was human.
>
> Alan Turing

After the war, the National Physical Laboratory on the outskirts of London invited Turing to join a computing project and in early 1946 they started work on ACE, the Automatic Computing Engine. By October 1947, the project was getting nowhere and Turing returned to Cambridge. A few months later Turing was given another opportunity when a second computing project started up in Manchester, and June 1948 saw the first practical demonstration of Turing's computer principle. The world had changed.

Computing

In 1965 Gordon E. Moore, a research director at the Fairchild Camera and Instrument Group, predicted that computing power would double every two years. He was right, and this exponential growth has changed the developed world.

Since the Babylonians used the abacus in the fourth century BC, humans have tried to find machines that can augment our ability to think. This started out with mechanical ways of adding numbers together. By replacing the complex Roman notation of numbers in the eighth and ninth centuries with Arabic numerals, European mathematicians made the process a lot easier. The advantage of the Arabic system is that it introduced a way of writing zero, and created fixed places for ten, hundreds, thousands, etc.

In 1623, German professor Wilhelm Schickard (1592–1635) built the world's first mechanical computer. It could work with six digits, and carry digits across columns, but he never got around to developing it particularly far. Frenchman Blaise Pascal (1623–1662) also tried, and in 1642 showed his eight-digit calculator. It worked but easily jammed. Charles Babbage's (1791–1871) massive steam-powered mechanical calculator never got off the drawing board, but this Englishman's punch-card-using "Analytical Engine" was described as a machine that could weave algebraic patterns just as mechanical looms weave flowers and leaves. It was still very basic, but these pioneers demonstrated that a machine could manipulate numbers.

As so often happens in science and technology, great advances occur when new ideas from different areas merge. This has certainly been the case with computing. In the 1940s people started using the newly developed electronic valves to drive calculating machines that operated using recently thought up Boolean logic, which asks a series of yes / no questions. By 1946 the University of Pennsylvania's Electronic Numerical Integrator Analyzor and Computer 1 had some 20,000 valves.

Valves were primitive electronic switches and were difficult to maintain, so Bell Telephone Laboratories' 1947 announcement that they had developed a transistor was a great leap forward. These were vastly smaller than valves, needed less power to run, and were much more reliable.

It wasn't, however, until Texas Instruments and Fairchild Semiconductor produced their first integrated circuit in 1959 that the world got a glimpse of the new future. All of a sudden thousands of electronic switches could be packed into a tiny space. What has followed since is the development of high-precision manufacturing technology, coupled with a greater understanding of the physical properties of the materials used, that has allowed circuits to become ever smaller and more complex.

The result is that computing power is now built into items from disposable musical birthday cards to massive computing systems that attempt to predict weather patterns. Since the 1970s such computing power has doubled every two years. The question now is whether this trend can carry on forever.

Claude Elwood Shannon

Some things are predictable and others are not. Claude Shannon was excited by the unpredictable aspects of existence, and called this "information." His solution to handling information created the dawn of electronic communication.

Born: 1916, Petoskey, Michigan
Education: University of Michigan
Major achievement: Created information theory that underpins computers
Died: 2001, Medford, Massachusetts

Take a die and write 6 on each face. Throw it and the answer is 6. Throw it one hundred times and the answer is six every time. You could record this by writing one hundred sixes, or more simply, 100 x 6. This led Shannon to realize that information was the unpredictable element of data. You can condense the hundred singly written sixes because the sequence is predictable — there is no information in the list.

Similarly, you could communicate numbers like π by simply sending the algorithm to calculate it, which is a lot simpler than sending the thousands of digits that have now been calculated for it. Once you have the algorithm, the number is predictable and manageable. The algorithm gives the information.

But spotting and condensing information is just a tool. In his paper *A Mathematical Theory of Communication* Shannon not only introduced the concept of information but also of the concept of communication theory. "The fundamental problem of communication," he wrote, "is that of reproducing at one point either exactly or approximately a message selected at another point." In other words, how to get precise information from one place to another without either losing vital elements or introducing extraneous and potentially erroneous additions that render the message inaccurate.

Thus, communication theory is concerned with how to transmit this information and what might go wrong with the signal during transmission that could cause the receiving person to misunderstand it.

Shannon further realized that you could code the results of flicking a coin with a series of 0s and 1s. More complex information like letters of the alphabet could still be coded this way, but each character would need five 0s and 1s. For example "a" could be 00001, "e" would be 00101, and "z" would be 11010. He also realized that the four-letter code used in the DNA that makes up chromosomes in cells operates in many ways like a four-value version of this binary system. Each of the 0s and 1s he referred to as a "bit" of information. Communication channels, he said, have a "capacity" that is measured in bits per second.

> I visualize a time when we will be to robots what dogs are to humans, and I'm rooting for the machines.
>
> Claude E. Shannon

Once he had turned information into bits, it was a short step to see that these could be represented by switches, with "1" for "on" and "0" for "off." Using these switches along with Boolean algebra, which he had studied during his first degree, Shannon showed that information could be processed automatically by electrical circuits. This has become the backbone concept in electronic circuits and computers.

Shannon's information theories eventually saw application in a number of disciplines in which language is a factor, including linguistics, phonetics, psychology, and cryptography. His theories also became a cornerstone of the early work in the field of artificial intelligence, and in 1956 he was instrumental in the first major developments in artificial intelligence research.

Index

For main scientist entries see contents pages. References to scientists are given only where mentioned other than their main entry.